Accidental Ranger

LYNDEL MEIKLE

2021
MOUNTAIN PRESS PUBLISHING COMPANY
MISSOULA, MONTANA

Too Many Good People

Where do I start saying "thank you?" Where do I stop? I owe so much to so many people and my desire to acknowledge them all is warring with my fear of leaving anyone out. Do I only thank those who have been a positive influence or do I owe thanks to people who drove me to take an unexpected path to avoid taking their path?

I have been helped, encouraged and healed by wonderful Park Service friends, cronies at the Montana Historical Society, FFA and Ag-Ed students and their teacher at Powell County High School, visitors to the ranch, editors, Forest Service allies, fellow members of Humanities Montana, the Vet Center in Missoula, brother Buck and sister Jeanie and their families, and all my teachers from the Café College. I owe thanks to the people who founded and created the history of Grant-Kohrs Ranch.

From my heart, my deepest gratitude to all of you.

Copyright © Lyndel Meikle
All rights reserved

Illustrations by Lyndel Meikle

Library of Congress Cataloging-in-Publication Data

Names: Meikle, Lyndel, 1946- author, illustrator.
Title: Accidental ranger / Lyndel Meikle ; illustrations by Lyndel Meikle.
Description: Missoula, Montana : Mountain Press Publishing Company, 2021. |
Identifiers: LCCN 2020045656 | ISBN 9780878427000 (paperback)
Subjects: LCSH: Meikle, Lyndel, 1946- | Park rangers—Montana—Grant-Kohrs
Ranch National Historic Site—Biography.
Classification: LCC SB481.6.M45 A3 2021 | DDC 333.78/3092 [B]—dc23
LC record available at https://lccn.loc.gov/2020045656

MP Mountain Press
PUBLISHING COMPANY
P.O. Box 2399 · Missoula, MT 59806 · 406-728-1900
800-234-5308 · info@mtnpress.com
www.mountain-press.com

Forward

An autobiography? Ridiculous!

When it was suggested to me, I laughed. I even imagined an appropriate title: "Nearly Always Almost."

Most of the things I've done sound more significant than they were.

OK. I lived in a one room cabin in Montana while my folks mined for lead and silver, but that was their story, not mine.

I was often in Vietnam during the height of the war, but serving coffee to men on military charters did not make me a combat veteran.

Firefighting? A couple of fires as line crew and a couple as security fell far short of my youthful dream of being a real Forest Ranger.

And then there was my audition for Ice Follies.

In fact, most of my life can best be compared to a patchwork quilt and a pinball. I did a little bit of many things and then bounced off to new experiences.

At last, however, I realized that I did have one tale to tell. It's the story of the education given to me by hands, hooves, claws and paws. My teachers ranged from several ants and a caterpillar to university professors, curious children and liars. Add roots to that list, both vegetal and historical, and you have nearly forty years on a ranch which is no less mine for also being yours: Grant-Kohrs Ranch National Historic Site in Deer Lodge, Montana.

Dan b. View out north window of VC, with jackleg fence, railroad, ranch house and Western hills There should be black and rusty streaks under the window where rain leaks in, but it's nice to be able to clean up such imperfections by just not "mentioning" them.

I wonder how many thousands of times I've looked out this window - at snow and at spring green, at magpies, horses, cattle, ground squirrels, deer, fox, coyotes, squadrons of geese and pelicans. A tiny slice of view - but a moving picture.

ACCIDENTAL RANGER

Exiled from Montana

A small girl stood in the center of the nearly empty room. Her fists were clenched and she frowned ferociously.

"I'll be back!" she proclaimed, as she was taken to the car which would carry her away from Montana.

"I'll be back!" she resolved, as the family began a new life in California.

"I'll be back!" she assured herself as years passed and Montana became a magical memory, kept alive by the two silver dollars she had saved as talismans, by the poems she wrote about her beloved mountains and by the fantasy world she kept within her.

"I'll be back!" she had promised, and one wonderful day—she was.

The story might be more dramatic if she had been a complete exile for the score of years which passed before she moved home, but there had been visits.

As a 12-year-old in fragrant orange "Tangee" lipstick, she sat high in a weeping birch tree in Helena and received her first kiss—memorable mainly because of the beauty of the tree. In her early twenties, she returned to seek out the cabin her parents had built across from their mine up Telegraph Creek and nearly despaired when she took the wrong fork and ended up on the Little Blackfoot River. She had believed herself incapable of forgetting the smallest detail of her Montana years.

*HUBCAMP ACROSS FROM THE MINE. OUR ONE ROOM
HOME, WITH MY BROTHER BUCK BESIDE ME.*

Soon, however, she found the right road, and there was the cabin, nestled in trees which were thicker now. The mine across the road was beginning to be hidden by the encroaching forest. Down past the outhouse and next to the creek was an ancient water wheel, built decades before she was born, and if she didn't remember it as she felt she should, she was quite willing to put it into her memory ex post facto.

It's cumbersome to write about the girl as if she were a stranger, so I will confess that I am the stubborn child who never let go of Montana. Or perhaps it never let go of me.

The most remarkable thing is that when I moved home, it wasn't like I remembered.

It was better.

Oddly, it was the airline which set my feet back on the trail to Montana.

———•••———

The years of exile can be dealt with quickly. I stayed in the west, in California and Oregon. I went to school some, learned a little, ignored a lot and eventually graduated from high school. College and I parted company quickly and I went to work—in an ad agency, as a stewardess, a store clerk, an insurance manager, ice skating instructor and a few other jobs before starting a lifelong career as a national park ranger.

Flying Tiger Lines was busy in the late 1960s ferrying troops to Vietnam, and I was a platinum-haired stewardess, more concerned with not wobbling in my high heels than with the moral questions of the Vietnam War.

Perhaps my most important contribution to the war effort occurred on February 18, 1968. That night we were on the ground in Bien Hoa, and, as usual, I was bagging up leftover milk and cookies from the flight in. Fresh milk, in particular, was very popular with the men on base. Suddenly the lights went out all over. A thunder of feet down the ramp revealed that the cockpit crew had run for cover, forgetting to guide us to the bunkers as we had been trained.

We left the aircraft and wandered down the tarmac, eventually finding a bunker. My contribution? I brought the milk and cookies with me.

I wasn't much of a mathematician, but after the Tet offensive in 1968, even I could not ignore the implications of full planes flying to Vietnam and half-empty planes coming out.

I started spending my days off hiding in the Sierra backcountry, trying to escape the inescapable realities of my job.

I dreaded the next time I'd have to stand at the door of the aircraft and say, "See you in thirteen months!" to soldiers who would file down the ramp into the sticky Vietnam heat. I wasn't going to see them in thirteen months. I knew it, and from their grim faces, they knew it too.

One flight stands out in my memory. Our crew picked up the flight in Japan, and the crew we relieved said the passengers—all Marines— had been singing since they left the States. They continued to sing as we flew to Vietnam, and as we began our descent they sang "America the Beautiful." I couldn't say goodbye. I shut myself in the cockpit and cried.

One final escape into the backcountry of Yosemite National Park decided me. I hiked out of the high country and phoned my boss from Tuolumne Meadows. "I quit."

Down in Yosemite Valley, I found work—initially for Yosemite Park and Curry Company, which ran most of the concessions. The job kept me in Yosemite Valley, which was where I wanted to be.

In my spare time, I started volunteering for the Park Service in their "VIP" program; Volunteers in Parks. In 1972, the interpretive supervisor told me they'd lost my volunteer application and gave me another form to fill out. I did so, without paying it much attention. Along about May I said, "I wish I'd applied for a seasonal job!" He replied that I had; he had given me a seasonal application, knowing me better than I knew myself.

Those were odd times for the Park Service. In theory, there were no women in law enforcement. However, I was afraid of guns, so I asked the range master if he would teach me to shoot. As it happened, I scored well enough that I "qualified," and started attending briefings and doing road patrol outside my regular duty hours.

One of my most vivid memories is of a patrol partner, his hands clenched on the wheel and his voice tense, saying (not asking) "I just don't know why a woman would want to do this job!"

Although both men and women led walks and talks, only the men were allowed to wear the uniform when I started volunteering. That was the decade of "Women's Lib" and we began creating a quasi-uniform of green jeans and white shirts, with the NPS arrowhead sewn on the shoulder. That wasn't enough for us; however, there was a uniform store in nearby Merced. I wish I could recall who instigated it and exactly how it came about, but we marched in and told the proprietors we were going to buy the men's uniform. They didn't seem to mind. Then we marched into the boss's office and said we were going to wear it. He simply said, "Well, take a picture and send it to Washington." And that was almost that.

There were a couple of problems when they finally produced a specific woman's uniform. For one thing, the Class A pants had no back pockets. "Where are we supposed to carry our wallets?" we asked. "It breaks up the line," was the ludicrous reply, as if female rangers were out there to model our backsides. For a while they scaled down the pockets on the uniform shirts, which meant that they were too small for our government notepads and our pens had an embarrassing way of working up and out.

Another problem with the women's dress pants was the height of the waistbands. Two or three inches higher than men's pants, it brought our gunbelts up so high the gun butt was tucked awkwardly in our armpits. The solution was to wear the field pants—identical for men and women.

MARILYN HOFF AND MYSELF IN THE PHOTO THAT WENT TO WASHINGTON, D.C.

By 1974, I was dressed in a Civil War uniform, sword at my side, working at Fort Point underneath the Golden Gate Bridge in San Francisco. It was part of the recently created Golden Gate National Recreation Area. Although there actually had been a few women who wore the Union uniform during the Civil War, I was careful to point out to visitors that I was wearing what the men had worn at Fort Point, and as far as we knew, there had been no women there. Charlie Hawkins, our boss, took a lot of flak for putting a female in that uniform. He was a retired army sergeant and I would have expected him to agree with the critics, but his response delights me to this day. "Just because we discriminated against women back then," he stated flatly, "is no reason to do it today."

Next, I "did time" at Alcatraz (also part of the park system after closing as a Federal Prison), but as I cast about desperately for a way out, I saw a vacancy announcement for a new park in Deer Lodge, Montana. Grant-Kohrs Ranch National Historic Site would commemorate the days of the open range. This era, filled with adventurous cowboys, wide-open spaces and fortunes made and lost created much of our self-image as Americans.

I could barely remember Deer Lodge. My brother, Buck, had been a young Golden Gloves boxer and we would occasionally go to bouts in Deer Lodge. That was about the extent of my recollections, apart from the fact that the town had electricity and flush toilets—a luxury by Telegraph Creek standards, although already known to me from part of each year spent in Helena. The city was also the home of Cinderella's castle, or so I had believed when I was little. The stone walls and brick towers of Montana State Prison had seemed to belong to the realm of children's stories. But I knew Deer Lodge was less than an hour from our old cabin and I decided that Grant-Kohrs Ranch would be my ticket home.

A quick trip to the new park in 1975 acquainted me with the ranch and the staff. Back in San Francisco, I wrote a groveling letter to the superintendent, explaining how much more it would mean to me than to anyone else in the world to work there.

The next year, I was driving a stuffed-to-the-gills sedan down the gravel road that led to the "house with taillights" I would call home for over a quarter of a century, and where I would continue to work even after a semi hauled that home away. It was a single-wide trailer, and certainly didn't fit in with the historic structures nearby. Surprisingly, visitors didn't seem to see it. It didn't belong, so they simply ignored it.

GRANT-KOHRS RANCH HOUSE.

PUTTING UP HAY WITH A BUCKRAKE.

The Grant-Kohrs Ranch Cows

Cows really get in the way when you're trying to be a cowgirl. There are also gates to open and close, followed by more gates and more gates.

Even when you and your horse have successfully negotiated them all and finally found a long stretch of road on which theoretically, you can gallop to your heart's content, the odds are that some large and stupid bovine will be blocking the way.

Actually, my trusty horse, Flash, was a little old for galloping, but that didn't stop me from resenting the cows.

There I was, living my childhood dream of being either a ranger or a cowboy (it had been a shock at about age four to learn I'd have to be a cowgirl) and it wasn't going according to script.

From the back porch of the ranch house I could count 44 gates. The 216 acres which surrounded me included 11 miles of fence.

And then there were those darned cows.

They could be peacefully bedded down a quarter of a mile away and I'd decide to leave a gate open since I'd be right back. Those cows must have been able to teleport. For by the time I returned, every single one would have nipped through the open gate and begun depredations on the flower garden or the vegetable garden.

If it was late afternoon and the train was about to roar past, they'd all be up along the track.

They could be up to their ears in good grass, but as soon as they had a bellyful, they'd break out and wander somewhere they didn't belong. The calves were cute but not friendly, and the by-products of their mothers' digestion made heedless walking hazardous.

It took me years to really "know" cows, and the more I knew, the more I liked them. But in my first year or so at the ranch, I was as uncomprehending as if I'd come from another world, which, in fact, I had. What, for goodness sake, did cows have to do with anything?

I had grown up on Roy Rogers and Gene Autry. Cowboys caught bank robbers and stage robbers. They thwarted bad guys who were trying to take over the ranch because it concealed a gold mine or, once the Cold War was in full swing, a uranium deposit.

Cows were bit players, necessary to provide a backdrop when the good guys captured the rustlers or saved the little child from a stampede.

An invitation to assist at a branding at a neighbor's ranch did not, initially, do much to educate me. Burning hair smells horrible. I hadn't bothered to understand why people still branded and I was mortified to realize that my self-image as a Real Montanan wasn't being borne out by my knowledge or skills.

Yet it was that branding which provided my first moment of enlightenment about what it means to raise cattle, what it means to be part of the rhythm of seasons which mean more to the heart of agriculture than mere market reports. I began to understand what it means to be a Montanan as I reestablished my citizenship.

I had started the day in the futile position of gate guarder. My role was to stand in the middle of a 20-foot gap and convince panicked calves that they did not want to dash past me and rejoin their mothers. I was quite bad at it.

Eventually I started working with the actual branding crew. A calf would be roped and thrown. Then I would hold one of its front feet and kneel on its neck while someone else held one back leg, another branded it and yet another doctored it.

CONRAD KOHRS WARREN BRANDING CATTLE IN 1930S. HE PREFERRED BRANDING IN A CHUTE, AS ROPING THE CALVES AND DRAGGING THEM TO THE FIRE PUT TOO MUCH STRESS ON CALVES AND THEIR MOTHERS.

9

At noon, we all trooped down to the ranch house for dinner, eaten in near silence. A mumbled "thank you" to the rancher's wife preceded an orderly shuffle out the door and then we were back to work.

As a treat, I supposed, I was allowed to switch from neck-kneeling to holding the back leg.

I was gritty and determined, but it wasn't enough. As soon as the hot iron touched the calf's hip, it yanked its leg away, bolted up and disappeared, half branded.

Embarrassed, I said to the rancher, "I've just started lifting weights." "Don't lift weights!" he replied. "Work."

Work

Work? I thought I'd been working for several years. Some had been office work, but I'd also done my share of physical labor; splitting firewood, holding up one end of a litter to carry out an injured hiker, putting in long hours and occasionally even getting a little dirty.

Soon, though, I found there was a different quality to ranch work. Some of the skills could be learned quickly but a deeper understanding of the tasks could only be gained over a long period of time—perhaps a lifetime. It was like learning the alphabet, and then discovering that those individual letters could be put together into words and the words could be strung together into stories and songs and poetry and other languages. It was endless.

Fortunately, one government negotiating with two competing railroads had not been in full agreement over who would take the blame if a train fell off a trestle over the proposed pedestrian underpass between the visitor center and the main ranch buildings. Their deliberations delayed our official opening for more than a year and my first summer was spent more on maintenance than ranger programs. I found I had a lot to learn about work.

We were building a jackleg fence and I was teamed with someone even more inept than I. At least I usually only damaged myself. My partner looked as if he were rehearsing a Three Stooges routine, with him taking all three roles.

Every once in a while, I'd smack myself with the heavy hammer as I drove in a 60-penny nail. Then I'd whine and snivel and make a fuss before getting back to work.

Then one day I glanced up just as one of the experienced fencers smacked himself with his hammer. He pressed his lips together and ducked his head for a moment, then went on with the job as if nothing had happened. Aha! He wasn't necessarily that much better with a hammer. He just didn't whine and snivel.

I tried the technique and within a few days was being teamed occasionally with the old hands, while we "shared" the disaster-prone member of our crew.

Over the course of the summer, our jackleg fences crept across pastures, along roads and through the willows down by the river. It did not occur to me to wonder why this type of fence was being built. I figured we were just replacing what had been there historically. In a way, that was true, but it had been used historically for several good reasons.

Lodgepole pine thrives in this arid country. Their small size made them perfect to support the leather lodges of the Plains tribes, hence the name "lodgepole," and they were equally useful for pole fences.

The ranch has good soil for growing hay, but the ground is very rocky. Jacklegs sit on top of the ground, so except for gates and corner posts, we had very few postholes to try to dig through the softball-sized stones.

Where the water table was high, sunken posts would have rotted off. Jacklegs, not being buried in the wet soil, didn't rot. They could practically be pontooned across boggy areas, and I was informed (but did not confirm) that the bottom rail of a jackleg fence is known as the "cowboy toilet."

I learned of one historic technique which we did not attempt to copy. In the 1930s, Skookum Joe Yankovich built fence at the ranch, and at the bottom of each posthole he dropped a scoop of arsenic. This was gradually drawn up through the posts by capillary action and prevented rot.

That summer was the merest introduction to fence building and only the start of my introduction to a new world of work.

It's work which is never really finished and which ensures the welfare and safety of people and animals alike. It's work which, if done sloppily or incorrectly, can lead to any number of problems, such as lost, stolen or injured livestock, injured people and—possibly most dangerous of all—the neighbor's purebred Angus cows encountering your romantically-inclined, escaped Longhorn bull.

"Good fences," wrote the poet Robert Frost, "make good neighbors."

Learning

There is what might be described as a Montana peculiarity when it comes to learning. Because it is considered rude to offer unsolicited advice, greenhorns are often left to their own devices.

This breeds a hardy type of independence and occasionally leads to injuries and embarrassment.

We were building a haystack with a huge mechanical pitchfork called an overshot stacker. The hay was cut with a mower pulled by a team of Belgian horses. Then the hay was left to dry for a couple of days before being raked into windrows—long mounds of hay. Next, the windrows were raked up and brought to the fork at the base of the overshot stacker. Finally, a team hitched to a long cable would raise the 20' fork and toss the hay into a pile.

Pete handed me a "people-sized" pitchfork and told me to groom the stack.

"What does that mean?" I shouted, as he drove off, but whether he heard me or not, I received no reply.

STACKING HAY WITH AN OVERSHOT STACKER, 1928.

After a while he came back with a load which was duly pitched to the top of a stack, already over 10 feet high. Some of the hay stayed on the stack, but some tumbled down the sloping sides. As Pete drove off for another load, I yelled, "Am I supposed to toss the loose stuff back up?" but perhaps the horses' harness was jingling too loudly, because once again there was no response.

I tried pitching the stray hay back up, but it was futile. The top of the stack was rounded and it just rolled back down.

Lounging by the fence was an old rancher who finally took pity on me. He strolled over and observed casually, "I used to wear myself to a frazzle up on top of the stack, trying to hollow it out before the next load arrived."

Then, with no further comment, he strolled off, leaving me to recover from my embarrassment, scramble to the top of the stack and frantically fork the hay out of the center of the mound. The next load fell into it as neatly as a scoop of ice cream into a bowl.

I was very new to the ranch at that time and was meeting people faster than I could really get a grip on who they were. It's odd now to realize that the nameless rancher who helped me must have become someone I knew well in the coming years.

At first, the older ranchers sort of blended together. They wore jeans and long-sleeved shirts, and if they owned a ranch and were over 60, the odds were they wore a narrow-brimmed gray Stetson and looked like they were somewhat down and out. They were often "stove up," or, as is frequently said in this part of the country, they looked like they'd been rode hard and put away wet.

The jeans, the narrow-brimmed Stetson, the long-sleeved shirts and the "hitch in their git-alongs" were all part of the world I had started to learn. Now, decades later, I have the jeans, the long-sleeved shirts and (after numerous ranch wrecks) an occasional hitch in my git-along, but I've never earned the narrow-brimmed Stetson. That's an honor beyond me.

But then, I'm still learning. I've heard that in other parts of the country, people aren't so reluctant to give unsolicited advice. Down South, it's said, you will not only receive the opinion of the person watching you, but you'll learn what your observer's cousin would do, and the cousin's best friend's sister's ex-boyfriend. Such help may make the learning easier, but hard-earned knowledge has a special value.

Some of the best lessons I've learned have come when I couldn't find anyone to teach me.

A Slight Conflict of Cultures

In many ways, a ranch run by the National Park Service is like any other ranch. When it's 40 below zero, it's just as cold for government cows as it is for any other cows. A 2:00 a.m. check of cattle during calving season is just as early in the morning for a government employee as for a private rancher.

One big difference, however, is a kind of detachment which stems from the knowledge that the people you are striving alongside today may transfer to Yellowstone National Park tomorrow, or the Statue of Liberty, or Death Valley.

Since 1976, several hundred people have been employed at the ranch. Some had scarcely arrived before they were packing up to leave. Some were seasonal workers who had no choice but to move on. Some stayed for years.

A few took the ranch to their hearts, and even after their careers had led them elsewhere, they remained part of the ranch.

Ed and Jean were here when I arrived. They both wore more than one hat. Ed provided ranch security and worked with the livestock. Jean cleaned, gardened, took care of museum objects, pampered every animal (wild and domestic) on the place, took care of stray children and taught me about keeping my temper.

It's hard now to remember why I was such an angry person when I came here, but I do remember that I never lost my temper appropriately. It was always some hapless passerby who took the brunt of my ill-humor. I can still get pretty grouchy, but at least now I'm ashamed of myself when I do.

Jean's voice was so gentle. She was as soft-spoken to a crabby person as she was to a tiny kitten or an angry cow. I remember my surprise when I was told she had trained K-9 dogs for the military. I pictured gentle German shepherds, winning over the enemy with friendliness and good manners.

She never embarrassed me when I'd make a mistake or a particularly ill-informed observation. She always tried to find a way for me to save face. My efforts to do the same for others are often clumsy, but she provided me with a standard to live up to.

Ed had taught vocational agriculture, and several times over the years, former students of his passed through and told me what an influence he had been in their lives. I work with local Ag-Ed students at Powell

County High School each spring, and wonder what I might do to be as useful to my students as he clearly was to his.

The two were quite a couple: Ed with his guitar and cowboy songs, Jean with such a sly sense of humor that you had to be alert not to be taken in by her occasional pranks.

Between them, they provided a refuge for a lot of local children who—though not really orphans—were as bereft of parental guidance as if they were.

One winter night I was sitting at my kitchen table, facing a window which looked half a mile across a pasture to the nearest light.

Suddenly I saw a shape flit past. Then another and another. Startled and a little frightened, I flung open the kitchen door and yelled angrily, "What's going on out there?!"

After a brief pause, several quavering voices sang, "Silent night, holy night…" It was Jean's Girl Scout troop, caroling for the "lone ranger," who was learning (with Jean's help) that though one might live alone

ED AND JEAN GRIGGS AND SMOKEY.

in the country, one wasn't really alone. You might think you're living in isolation, but if you ever need a hand, folks are there so quickly you might wonder if they'd been standing beside you the whole time.

Ed and Jean eventually moved to Oregon. Other employees took their jobs, but no one could truly take their places.

Cats and Dogs

Everybody knows about cats and dogs. Dogs are loyal, friendly, protective, obedient—sort of a canine version of a Cub Scout with ambitions to become an Eagle Scout. Cats are sneaky, independent and possibly possessed by evil spirits.

Unfortunately, when I arrived at the ranch, I was informed that I could not have a dog.

Equally unfortunately, my single-wide government trailer was infested with mice, the previous resident having inexplicably stored fifty-pound sacks of rice and flour in a bedroom closet.

There were quite a few barn cats lurking about. Ed and Jean had an exceptionally intelligent dog named Smokey who knew the difference between strangers coming in the daytime (welcome), strangers coming in at night (unwelcome), wagons which were meant to be used (hop on in) and wagons which were part of the amazing collection of artifacts preserved at the ranch (paws off).

He was a good companion for me, but he had one serious flaw. He liked cats. The cats also liked him. They palled around together, curled up with each other in cold weather and on one incredible occasion I saw Smokey wading in the shallows of the Clark Fork River with Victoria, the calico cat, wading along behind him.

I gradually got used to the cats being around and relaxed a bit when I found they weren't constantly plotting to slash me.

Meanwhile, the mice had a field day in my trailer.

I set traps and caught thirty-six in one week. At that point, I opened the front door, called out to the nearest cat (the aforementioned Victoria) and asked, "Would you like to come in?"

She caught all the remaining mice and then she trapped me. Before I quite knew what had happened, I was involved in a nightly ritual of taking the cats for a walk. Victoria was joined by Puff (so named because he was the same color as the dog; a Puff of Smoke) and Fred (of the hairless tail).

One day a new cat appeared. He was a shorthaired orange tom with stripes. I named him Beechnut after a gum advertisement of that day which ran, "Yipes! Stripes! Beechnut's got 'em."

I wasn't yet sufficiently fond of cats to worry about their fights as they established the new pecking order to accommodate the newcomer. By the time I was a fanatical cat lover, I knew better than to interfere, or so I thought.

Gradually, however, I learned that a couple of sharp claps and a firm, "No fighting, no biting," would break up nearly every fight among the regulars, even when they were outside and I was issuing the orders through an open window.

I also learned they would come when I whistled. They didn't always want to, and they sometimes didn't stay, but when I wanted to check that they were all well and accounted for, they would at least come within a few yards.

They taught me tricks. That is to say, close observation of their behavior taught me tricks. I noticed that every time Pierre approached me, she would drop and roll when she was about ten feet away. So,

BEECHNUT, THE LAST REAL COWCAT.

17

when she got about twelve feet away, I'd say, "Hey, Pierre, drop and roll!" and she—in apparent obedience—would do so. "How did you teach her that?" people would exclaim. I'd just smile deprecatingly and shrug my shoulders.

I built a "cat window" through which they could come and go, and they became official residents. A cat door would have been unwise, since skunks were regular visitors on the front porch, but skunks don't jump. Raccoons were another matter, but it was some years before they invaded.

The cats had nearly won me over even before the night I lay in bed, groaning slightly from the pain in a foot which had just had a broken bone cut out of it. They curled up with me for a while, but as I moved restlessly, they got up and I heard them diving out their window.

"Humph!" I thought, resentfully. "A dog would have stayed with me! A dog would have comforted me!"

Soon, I heard the thump of an incoming cat and a mumbled "mreow." Looking over the edge of the bed, I saw that Beechnut had brought me a mouse.

Another thump brought Fuzz Bucket in with another mouse, followed by John Muir (pronounced Me-you-er). The strict rule in the house was "No mice past the kitchen doormat," but they seemed to know this was a special circumstance.

They looked at me expectantly. Then when I didn't immediately devour the offerings, they bolted back out and returned shortly with more.

MY FAVORITE CRITIQUE FROM A MEDIA PERSON WHO TOLD ME, "YOU OBVIOUSLY HAVE NO ARTISTIC ABILITY WHATSOEVER, BUT YOUR DRAWINGS ENHANCE YOUR COLUMNS." UH . . . THANKS?

I couldn't get up, but fortunately there was an empty can within reach. I plopped the mice inside and popped the lid on it.

Seeing the mice disappear, the cats seemed satisfied. They had saved me from starvation. They piled back on the bed, nestled down and were soon asleep. Years passed and Fuzz Bucket turned Pierre, John Muir became Trout. Cats of every size, color and personality passed through the window into my home. Their home. Our home.

Although I'd like to make my bed
A furry little sleepyhead
Is curled up cozy, soft and warm
An unmade bed will do no harm.

The Wild West?

When I had come up to check out the newly-created Grant-Kohrs Ranch in 1975, the town was as western as a town could get. There were horses everywhere. One dashing young cowboy was cantering along the frontage road, holding the lead ropes of three other horses. What a place! There were more horses than riders!

Everyone seemed to be in western clothes. Even the waitresses at the local coffee shop looked as if they were dressed for a brisk gallop.

Less than a year later, I moved to the ranch and was too immersed in learning my new job to notice the change that had taken place.

The waitresses wore ordinary skirts and blouses. There were horses to be seen in pastures, but it was rare that I saw anyone riding them. Young kids I'd seen in Stetsons, white shirts and string ties now wore t-shirts and jeans like kids almost everywhere.

It was probably a year or more before the light came on: My reconnaissance visit had been made during the Tri-County Fair. The plethora of cowboys had been there for the rodeo. The Stetsoned and string-tied young kids had been dressed up to show their 4-H livestock.

The Ranch Visitor Center is directly across from the fairground and learning about county fairs became an important part of my ranching education.

I watched the 4-H hog judging, impressed by the seriousness of the contestants. Then it became more serious.

They'd done a good job raising their hogs. The animals were in good condition and comparatively well-behaved, though I had to repress a giggle as one enormous animal decided to take a scenic tour around the arena, dragging its exasperated small owner along with it.

The ribbons had nearly all been awarded when the judge gave the youngsters a stern look. It was time to give out the showmanship award, but there would be no such award that year. He solemnly informed the contestants that none of them had measured up to the standards required.

I was astounded. I was accustomed to a world where people were so careful not to hurt anyone's delicate feelings that blatant misdeeds were excused. Here in Deer Lodge, "mere" children were expected to live up to certain standards and take the consequences if they didn't. What a concept.

I wasn't around for the auction of hogs, cattle, sheep and various small animals. I missed the heartbreak of some of the youngest children as animals they'd raised, lovingly groomed, named, trained and dedicated the past year to were sold. And not sold as pets.

It was a hard lesson, and one I wish I'd learned when I was a kid. It was a lesson I finally had to learn when the bottle calves I cared for at the ranch grew up and either became part of the breeding herd or went the way of all beef.

The odd thing is that it doesn't make you callous. The close connection you establish with a particular calf helps you understand all the cattle. All the behaviors and needs and characteristics of your special "pet" make you aware of the rest of the livestock and the respect you have for an animal you know so well gradually extends to the animals you don't work with as closely.

If I were a kid again, I'd be proud to raise a 4-H calf. I'd feed it and groom it and show it, and I'd do my darndest to get the showmanship award.

Then, when it was auctioned off, I'd probably sniffle some before dedicating myself to the next one.

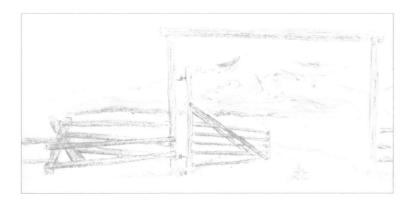

Gates

It's an old "truth" in ranching country: The man drives and the woman gets the gates.

This dates back at least to the days when trucks did not have power steering or maybe even to the time when a team of restive horses had to be reined in.

I suspect, however, that in the days of actual horsepower, the man simply handed the lines to the woman and got the gate himself. Climbing down from the high seat of a typical wagon in a long dress would have been difficult. So, it was probably with the advent of motor vehicles that it came to be considered easier for a woman to get the gates than to drive.

Trucks are easy to drive these days, but traditions die hard in cow country.

This was made clear to me by Ole, who drove me all over the countryside when I started to work at the ranch. We surged up and plummeted down slopes which terrified me then. It seemed we were always hurtling over cliffs and driving up grades so steep I expected the truck to flip over backwards. And every few minutes we'd lurch to a halt while I jumped out of the truck, opened a gate, Ole drove through, I closed the gate and jumped back in the truck.

After all these years, I now drive those same routes and occasionally wonder why I found them so intimidating. Slopes no longer hold any terrors for me, but I still dread the gates which seem to leap in front of the truck every acre or so.

There are many kinds of gates on a ranch. There are metal, wood and wire gates. There are gates which do a reasonable job of keeping livestock in and gates which seem to open like magic to some equine or bovine "open sesame" and let the whole herd out onto the railroad tracks, generally when a train is due.

But in another sense, there are only two kinds of gates: Those you can open, and those which require dynamite, a weightlifter or a tank to get through.

The most evil are barbed wire gates. These are generally made of three or four strands of barbed wire, strung horizontally about a foot apart. They are firmly attached to a stationary post on one side of the road and to a moveable post on the other side.

One of the biggest problems with barbed wire gates is that they are typically strung in the summer when the wire, being warm, has expanded. Then you have to open them in the winter when the wire, being cold, has contracted. This means struggling to lift off the tight loop of wire which holds the top of the gate against the adjacent fencepost. You accomplish this by squeezing the moveable post against the fencepost in a gasping sort of bear hug until the loop has enough slack to lift it off.

Terrible things can happen when you try to open or close one of these gates. Because they're made of barbed wire, the odds are you'll be poked with a barb or two as you wrestle with the wretched thing.

BARBED WIRE GATE: NOT, AS MY MOTHER WOULD HAVE SAID, WORTH THE POWDER TO BLOW IT.

It's hard to describe the exact manner by which you can find yourself with the sleeve or shoulder of your jacket or a long strand of your hair caught to the post inside the restraining loop, but trust me, it happens. And if you've used your last strength to squeeze the posts together so you can drop the loop back down to close the gate, you have no strength left to extricate yourself.

Meanwhile, the driver may or may not be snickering, but if you're stuck long enough, he may step out of the truck (with an air of condescension) to help you.

So, every time we were going over to the west side of the river, where there were several barbed wire gates, I groaned inwardly, having learned that groaning outwardly is a waste of time in the world of agriculture.

Then one day, Guy and I were going to the west side. I glanced down at the desk, and there were the keys to the truck.

So, I took the keys.

I drove and Guy got the gates. He didn't complain about the switch, either.

That wasn't the end of my gate-opening days. We were never so uncouth as to actually fight for the keys. Certainly not! However, once it was decided that we were going somewhere which involved barbed wire gates, we seemed to get a move on more expeditiously than before because whoever got to the keys first, drove. It came to be accepted that neither gender had an absolute right to the driver's seat.

At first, I thought I had won some sort of victory over the male of the species, but it finally sank in that men weren't actually making us get the gates. We were letting them drive.

The lesson was simple: Take the key.

Trucks—Especially Goldie

Speaking of truck keys, we had a lot of trucks over the years. I don't remember many of them. There was the red Willys jeep which had been a wildland fire truck. You had to pull the choke out to start it, and my most vivid memory is constant and quite unnecessary advice from bystanders to "choke it" whenever the engine didn't immediately roar into life.

I can only recall one truck which ever got a name: Goldie.

She was anything but new when we got her. Her cab was gold. That is, she was gold where minor collisions hadn't scraped her down to bare metal, or where an unexpected streak of blue paint from another vehicle or red paint from a five-bar wooden gate hadn't marred her golden perfection.

What a fantastic truck! You didn't even need to roll the window down to hear people speaking to you from outside, because there was a gap in the door frame wide enough to slip a cookie through.

It's just as well that you didn't need to roll the window down very often: The knob on the window crank was broken. That was a minor matter, and her perfections more than made up for her imperfections.

Of course, her upholstery was in pretty rough shape, but at least you could slide across the rips in the vinyl. Cloth seats rip up even more easily and make a non-slip surface which is very difficult to get across. It wasn't Goldie's fault that the seats were that way. We didn't always remember to take the pliers, screwdrivers, bolts, shears and other paraphernalia out of our back pockets before sliding in.

Her front bumper was a wreck, or perhaps I should say several wrecks. That was to her credit, really. If she hadn't been built solidly enough to take the abuse, it would have been passed on to us.

Don't get the idea that we were reckless drivers, but a feed truck is an entirely different species from those sleek and shiny weekend pickups used for hauling home large quantities of paper towels and pet food from big box stores.

Chain saws, leaky oil cans, sliding metal tool boxes, firewood, fence poles, tons of hay, an occasional ailing animal and trash do little to enhance a truck's appearance. There is a difference between a ranch truck and a feed truck. A ranch truck retains road worthiness which enables one to drive it to town. A feed truck is generally refueled by a gas can at the ranch and rarely experiences the glamour of a paved road.

One of my earliest feed truck adventures was to drive across a bumpy pasture while Ole perched on the back bumper, ready to jump off and capture an elusive calf. I was terrified I would run over the calf, the anxious cow or Ole. Thirty years later and with a bit more confidence, I performed the same chore as our night calver jumped off, captured a calf, tossed it up on the truck bed, jumped up behind and doctored it while I attempted to outrun its homicidally enraged mother. Not a typical night's work, but necessary.

Besides being used as cutting horses, feed trucks occasionally get pressed into service as temporary gates. A high-speed positioning of trucks in inconvenient gaps is one reason for the numerous dents they sport.

GOLDIE.

Goldie's hood didn't latch. No problem. It made it easier to check fluid levels and belts, and although the sensation of speed seems great when chasing a calf or racing to plug a gap in the fence, speed is relative. The hood was unlikely to pop up at 20 miles an hour—two or three times the average off-road speed—and the safety catch still worked in any event.

For a while, the bottom hinge on her driver's side door was broken, but we just had to open it carefully and slip in through as narrow a gap as possible. It even got fixed!

When it came to speed, Goldie beat them all. You could put her in four-wheel low and she would creep along so slowly that you could get out, walk alongside her and feed out the hay bales on the back and even retie your bootlace without the least fear that she would get ahead of you. This gear is nicknamed "granny low." That seemed disrespectful to grandmothers, so I usually called it snail gear, or (if we were being formal) Honorable Grandmother Low.

I've never owned a vehicle with an automatic transmission, though I've had to drive a few. Junk! For ranch work, a stick shift is the only way to go. I will confess that it was hard to put Goldie's four-wheel drive in, since the stretch to the shift stick for the four-wheel drive was so far to the right that I nearly had to lie across the seat to reach it. On several occasions I straightened up from that stretch only to find that one of the ranch horses had stuck its head in through the driver's side window.

Goldie's gas gauge didn't work, so I soon learned to check the mileage since the last fill-up before we set out, since a long walk back on frozen, rutted roads in sub-zero temperatures was not an attractive prospect. It wasn't a bad habit to get into, even with a working gas gauge.

Surprisingly, her windshield had no cracks, but the driver's side mirror had a hairline crack near the top which leant a carnival sideshow air to winter feeding. The reflected heads of cows and horses coming alongside from the left would suddenly duplicate themselves in the fractured top of the mirror, while the unbroken lower part showed only one body.

It might be an exaggeration to say Goldie saved my life, but she certainly saved me from a bad wreck one winter morning. We had received an unusually heavy snowfall the night before and the predawn light was dim: The snow and the sky were the same color and the road was a bit hard to find. A drift had built a deceptively solid-looking cornice from the edge of the road out over empty air above the big irrigation ditch. Suddenly the road vanished from under the left front tire. The other three wheels took over as I frantically cut right, and the left front regained the road just as the left rear lost it. Three wheels at a time, we got across the chasm. It took seconds, but my nervous giggles lasted considerably longer.

So, let's take inventory: Dents (large and small), leaky door frames, terrible paint job, broken window cranks, torn upholstery, broken hood latch, cracked mirror, no gas gauge…oh, I forgot to mention that there was no inside door handle on the passenger side and that door couldn't always be opened from the outside so the passenger would have to slide across to enter and exit from the left.

But Goldie was a great feed truck. Why?

- If treated right, she always started.
- Her heater worked.
- Her low gear was perfection.
- She had a great radio.

For a feed truck, there is nothing more to ask.

Sounds

When you are driving a tractor, it's difficult to hear a word anyone says to you. It was natural, therefore, to assume that no one could hear me when I was driving the tractor.

I do not have a good singing voice. In a singing contest between me and a goat, the goat might lose, but I wouldn't bet on it.

So, the tractor provided me with an unprecedented opportunity to sing at the top of my lungs without offending anyone. I couldn't even hear myself.

This was early in my ranch career, and I had been given a fairly easy task by a man who had contracted to put up our hay.

The process of growing hay begins as soon as the ground thaws in late spring.

Some parts of the country get enough rain to enable the pastures to grow on their own, but this valley, like much of the West, gets only 10 inches of precipitation a year, counting melted snow.

Runoff from nearby Mount Powell provides us with much of our irrigating water. It is mainly flood irrigation, which uses the natural force of gravity to bring water from ditches on the high side of the pasture across the downward slope.

The Grant-Kohrs Ranch grows a mix of grasses mingled with a little alfalfa, which is not a grass but a legume. In mid-summer, when the seed-heads on the timothy grass in the south pasture are just starting to mature, the water is cut off to the ditches and the fields dry up. The hay is cut and dries for a couple more days before it is raked and stacked or baled.

That first year, it had rained right after the hay was cut. This meant it had to be turned over so it would dry clear through. Wet hay gets moldy and can even generate so much heat that it will spontaneously ignite.

I was entrusted with turning the hay with a wheel rake. As I circled the field, I sang everything from campfire songs to snatches of opera arias. The tractor was a classic little workhorse with a rather noisy engine.

I had just learned that cattle baron Conrad Kohrs had a brother in the meat packing industry who owned a company which became part of "Oscar Meyer," and I'm afraid I even belted out a couple of choruses of the Oscar Meyer wiener song, substituting "Henry Kohrs" for the current company name.

I was allowed to enjoy that vocal freedom for a while. It was, in fact, several weeks before I realized that, though the tractor driver can't hear anyone else, everyone in a quarter mile radius can hear the tractor driver. Thirty-five years later, I ran into the son of the owner of that tractor. He mentioned that he and his mother had just been laughing about my singing the day before.

Country acoustics take some getting used to.

"MY" FIRST TRACTOR.

Having transferred to Grant-Kohrs from Alcatraz (one of the stranger units of the national park system), I was used to the noises of San Francisco.

My first thought as I settled into ranch life was that it was wonderful to be in the peaceful, quiet countryside.

It was certainly true that sirens, jet aircraft, clanging trolleys, roaring buses and—in those days—teenagers carrying portable stereos the size of a hay bale were generally absent from the country.

On the other hand, just try to listen for an elusive, distant sound and I can nearly guarantee that a chorus of killdeer, snipe and magpies will burst forth.

Snipe, by the way, really do exist. They are also called the American woodcock. Because unsuspecting people have been lured into futile, nocturnal hunts for snipe, the poor bird has been relegated by much of society to the realm of myth.

It's an interesting bird. It makes an eerie sound that is known as winnowing. It soars aloft, then dives at reckless speed. Wind catching in its cupped tail feathers makes the odd warble. They fly higher than people are expecting, and by the time the sound arrives, they are elsewhere, so they can be hard to spot. Naturally, people ask, "What's that?" When you reply that it's a snipe, you are usually rewarded with a skeptical look and a knowing, "Sure it is…"

I wouldn't mind being disbelieved if I was telling a tall tale, but one of the most frustrating things I know of is to be doubted when I'm telling the absolute truth!

So, the country boasts voices that may or not be heard, depending on whether you are on or off the tractor. And there are small birds making a racket which could drown out a rock concert. To top it off, there are acoustic effects which would make a ventriloquist green with envy.

Fortunately, I discovered by overhearing someone else's distant conversation that distant voices float lightly across the yard below the ranch house to echo off the buggy shed wall and echo the words all the way over to the blacksmith shop as clearly as if the speaker was standing next to me.

Kids speaking a quarter of a mile away at the river might as well be just outside the door.

It's not only voices which echo that way. When I was in the barns and sheds, it was completely impossible to tell from which direction trains on the nearby track were coming. The sound came equally from all directions—even directions in which there were no tracks.

Sometimes it was sort of companionable. When I was chopping wood for the barrel stove in the blacksmith shop, there always seemed to be another woodchopper keeping up with me, just a building away.

Hearing things in the country is a little different from hearing things in the city.

I was to learn over a quarter of a century that the sounds you hear so clearly aren't always enough. You have to listen, and you have to think about what you're hearing.

Listening

What do you hear when you listen?

No one is really free from the tyranny of technology these days. Cell phones and other electronic intruders are as pervasive in the country as the city. But the ranch provides another communication system which can be quite entertaining and occasionally just as vital once you learn the language.

Most of my work has been giving interpretative programs to the public. For about 30 years, however, I held a law enforcement commission. There were two law enforcement issues in those early ranch years: poaching and vandalism. Some experienced poacher would work through the dense brush along the river and wait until just before sunset. A shot or two would be fired, then all would be silence. Evidence of a successful hunt would be removed in the dark.

Not only barn walls produced confusing echoes. On a cold day, the ridges and folds of the hills to the west could make a gunshot from the Montana State Prison pistol range sound as if it was coming from the nearby river, though the range was four miles away. Conversely, the sound of a gunshot on the far side of a nearby haystack was nearly inaudible.

People are easily confused. As we look frantically in every direction, all we really need to do is look at the horses and cattle. The horses, in particular, look in the proper direction for several seconds before deciding it has nothing to do with them and they can go back to grazing. When those elusive gunshots echoed, I learned to look at the horses.

Coyotes tend to stay on the west side of the river, most of the foxes on the east. For years, I'd hear the coyotes at night and picture a pack of them sitting together as they sang. One hot afternoon, however, I climbed the highest hill on the west side, hoping to catch a cooling breeze. It was so pleasant up there I stayed until nearly sunset. Then a coyote started to howl at the base of the hill. Another coyote joined it vocally but not physically. It was on the other side of the hill. Others joined in, from all different directions. I listened for a while, surrounded by their canine serenade, then wandered homeward. The chorus continued, but the farther I got from the vocalists, the more it sounded as if they were all in one spot. What I had always pictured as a pack had evidently been long-distance calling.

There are sounds you can ignore (magpies, geese, trucks on the freeway a mile away) and sounds which herald a problem. A cow isn't deliberately alerting you when she calls for her calf, but your subconscious hears it, and when the calling goes on longer than necessary for a simple, "Junior, I said get over here NOW!" you tend to notice, because it means your assistance may be needed to reunite the pair.

Dogs barking, chickens squawking, flocks of ducks being spooked along the river; the country is full of sounds. The weather affects how the sound carries and ranch work can drown out other noises. Gradually, your "hind brain" learns to sift through the noises and lets you know when your attention is required.

I have a bad habit of imitating moos, whinnies, clucks and various birdsongs. If I listen carefully I can occasionally imitate the sounds well enough to fool some of our loquacious critters. I like to trick meadowlarks into changing from one song to another. I only know one of their tunes perfectly, but if I whistle it in response to another meadowlark's trill, I can often get it to change its tune.

Owls are even more fun. Male and female great-horned owls have different calls. They alternate calls to each other, but I'll break up the monotony with random hoots until they lose track of who's "whoo."

It's a fairly harmless diversion, and as soon as I stop they get back on track.

The barn cats didn't consider it harmless when I'd start the coyotes howling, expressing their displeasure with sharp meows. They were probably thinking that issuing an invitation to a coyote was singularly stupid.

I only once managed to fool a raven, though its call seems fairly simple. Ravens are a lot bigger than I realized. This one, thinking it had found a friend, landed practically at my feet, looked up, aghast, and took to its impressively large wings.

There is one animal imitation that I would love to learn. Alas, I'm an abject failure at "blatting" like a frightened calf. If you want to gather the cattle in a hurry, just do a good "blat" and the entire herd will come running.

That's because the animals are listening too.

The Forge

There are a lot of ways to learn. Formal schooling, observation, trial and error all have their good and bad points. One of my lasting regrets is missing an opportunity to learn from a fantastic teacher because of my own foolish gender stereotyping.

Ironically, despite my often-strident woman's lib philosophy, I assumed blacksmithing was a man's job. At least I assumed the Park Service would insist on its being a man's job. They often did.

That doesn't mean I wasn't fascinated by it. Over the years, the ranch had hired several men to demonstrate blacksmithing. Most were highly skilled. Some were good interpreters, but they were often more interested in just doing the actual forge work than explaining it to visitors. Their philosophy seemed to be that people would understand the work just by watching.

I watched, and being a snoopy sort of person, I also asked a lot of questions. It helped that although a large part of my job was supposed to be law enforcement work, the ranch was mercifully free of crime. Watching the blacksmith at work was more interesting than looking for non-existent evildoers.

Then one day in 1986, our blacksmith abruptly left. We had school groups scheduled and I offered to fill in until we hired someone new. I could light the coal forge. I could get a bit of iron red hot and make some noise hitting it with the hammer. That would be enough for a couple of weeks.

Frank Westfall, the local Vo-Ag teacher, dropped by and offered to teach me, but I knew I'd only be in the shop a short time. After all, it was a man's job. Frank showed me how to sharpen and temper a mining pick, but I declined further instruction since I believed I'd not get a chance to put it to use.

Tempering, by the way, is the process of hardening the metal, and there are a lot of variations. A mining pick, for example, needed to have the surface made hard enough so it didn't dull quickly, but the core of the pick had to be less hard so it wouldn't fracture.

It proved unexpectedly difficult to fill the blacksmith/demonstrator job. Time passed and I found I wanted to do more than just fill in. For a couple of years, I bought the coal and steel I used, assuming that if the budget folks didn't get a supply request, they'd never focus on the fact that a female was "impersonating" a blacksmith. Times have changed somewhat.

I soon realized that trying to teach myself was stupid. I was trying to reinvent and rediscover everything blacksmiths had worked out since the beginning of the Iron Age.

Having declined Frank's generous offer, I was embarrassed to go back and ask for help. I turned to books. Some were useful. Some were not. Possibly the worst advice I've ever read came from a 1980s text which said, "You can wear safety glasses if you want, or you could just squint." (A close runner-up for worst advice was the pesticide label which detailed all the dreadful consequences of misuse and then said, "Warning! If you cannot read English, get someone to explain this to you.")

Most modern blacksmithing books were filled with artistic projects. Dragonhead fire pokers and twisty Colonial candle holders were all very well, but didn't have anything to do with ranch shop history. Blacksmithing books of the 1800s assumed you already had some grounding in work which was all new to me.

Fur trade era items had some tenuous relevance. Johnny Grant, the founder of the ranch in the mid-1800s, was the son of a Hudson's Bay Company trader. Fire strikers (the "steel" of flint and steel for fire lighting) were interesting and easy to make, but as relics of the fur trade era, they were also about a century out-of-period for our 1935 shop.

Another common demonstration in historic blacksmith shops was nail-making. It's fast, it doesn't use much material and the result can be given to a wide-eyed young visitor. I hate to think how many nails I made before I realized I was sending generations of children away with the idea that a blacksmith in the 1930s had nothing better to do than make nails. After all, machine cut nails were common in the 1800s.

Making horseshoes had the right "look," but making a horseshoe doesn't make one a horseshoer. It didn't matter how many times I explained that a blacksmith was an ironworker and a horseshoer was a farrier, as soon as I shaped a shoe, I started getting questions about horses and I had absolutely no desire to take up that trade. It's hard on the back, legs and hands, and even a good horse can hurt you.

Then I interviewed Neil Ruttenbur, who had worked on the hay crew at the ranch when he was a teenager in the 1950s. He told me that he and the other boys had played around with the forge, making twisted hooks which still hung on many outbuildings. Meanwhile, I'd been making a lot of hoof picks (for cleaning horses' feet) as a prop to explain the cowboy's work. A hoof pick with a spiral twist became my standard demonstration.

It may seem odd that a person could make hundreds of hoof picks each year without finding it boring, but the ironwork was just a bit of drama to focus people's attention. Besides, the "new" didn't come from making the pick. I got it from the visitors. There were people I learned from, people I taught and people with whom I got to share an adventure into a not-too-distant but very different past.

The ranch had a lot of visitors with farm and ranch backgrounds. By the time I was learning forge work, I'd embarrassed myself enough times to have learned not to pretend an expertise I didn't possess.

Whenever a visitor's comment revealed experience in a shop, I had no hesitation in asking about any and everything which was puzzling me. And there was a lot!

Besides the forge, there was a workbench covered with unfamiliar tools. In the back of the shop was the line shaft; a long overhead shaft, belt-driven by an electric motor. Grindstones, a drill press and a grinding wheel all ran off the same shaft.

The workbench drawers were also full of mysterious tools. I didn't even know what they were for, much less how to use them. Old catalogs gave me the names of the tools, but couldn't give me the skill to use them.

Color and noise can capture the visitor's attention, but we live in a world of color and noise. The shop needed to be more than just a primitive light show.

That's when a little historic research came in handy. We had a portable microfilm reader and reels of film of the Deer Lodge newspaper from the 1860s into the 20th century. I'd take it home at night, set it in my lap, tip back the recliner chair and read the news of the previous century.

Careful reading turned up the story of Giles Olin, Johnny Grant's blacksmith. A consummate craftsman, he had the misfortune to lose his legs to frostbite in 1864. That didn't slow him down, however. As soon as he recovered, he built his own wooden legs and proceeded to accomplish more in the next 12 years than many people manage in a lifetime, including building a commercial greenhouse, an ore concentrator, a skating rink, a silver smelter and a patented propulsion method for canal barges.

Jens Pederson also still was shoeing the ranch's draft horses at eighty-two. He was reputed to be the strongest man in Deer Lodge, the best bootlegger and he grew the prettiest gladiolus. Discreet questioning of local old-timers confirmed my suspicion that the leftover mash from his still was put on the garden. He must have had the happiest "glads" in the valley.

The characters and abilities of the old-time smiths became a large part of the program as I concluded that demonstrations weren't—and shouldn't be—about my blacksmithing skills (or lack of them) but should be a way of taking the visitors out of the present and putting them into the past with Giles and Jens. I hammered less and spoke more.

But there was something very basic missing when I tried to focus on their stories, and the irony finally hit me. I may have been right that the work was not self-explanatory, but it is not possible to convey a hands-on world with your hands off. The work still had to be seen and heard.

The kind of men who knew how to do farm shop work were not, generally, the kind of men who wrote books about it, and they were right. Some things are best-learned on the job. A book might describe "red hot iron," but it isn't red hot. It's a range from dark red through yellow-white, every shade of which means something different. And you need to understand that iron which appears to be only red hot in the late afternoon when the sunlight reaches across to the anvil is dangerously close to white hot in the usual gloom of the shop.

There's hot-rolled and cold-rolled steel. You need to know the difference and your hammer hand can tell you which you're working with, even when the shipping order says otherwise. Cold-rolled steel has been formed at a low temperature, "work-hardening" it. Despite being heated, it is very hard to shape.

A book can tell you that iron is black from 900° down, but unless you have been tired, angry and in a hurry and mistakenly picked up a bar of hot, black iron with your bare hand, you may never fully understand that fact. And once you have made that mistake, you'll never forget it. I won't.

I have learned that when craftsmen pick up a familiar tool, there is a knowledge and respect for that tool which is instantly recognizable. They take it up as they would if they were about to use it. A hammer will be grasped just where they would hold it, and they usually give it a shake or two to get the balance just right. I never saw one of them hold hammer up by the head. It would be like trying to show someone how to use a soup spoon while holding it by the bowl.

The trouble with most of those hands-on craftsmen is that they were going to pass their knowledge on to their sons—and their sons didn't want to learn. World War II came along and when it ended, the world had changed. Rural roots were yanked out of the soil as young men went off to war. They came home to an industrialized nation. There were new kinds of jobs and easier money and new ways of living and thinking.

They could throw broken things away and buy new ones. They could buy new things even when the old ones weren't broken. But we lost a lot for all we gained.

My coal forge is a thousand degrees hotter than most current gas forges. The fire is bigger and can be adjusted to heat iron with more precision than the often-diffuse heat of gas. When a local rancher needed a couple of large pieces of iron bent, he brought it to me, explained what he needed and it was the work of a few minutes. It didn't even require any shaping beyond the simple bends. His only alternative, he told me, was a $900 new part.

That was years ago. It was an important moment for me because it began my fascination with and understanding of the ethic of "repair."

It has been a long, gradual education. Luckily, it has also been fun and interesting. There is a snide and generally false saying, "Those who cannot do, teach." It has some relevance in my case, since I'm sort of a blacksmith by default. But after Frank Westfall passed away in 1990, it looked like blacksmithing would have to be dropped from the general curriculum of Ag-Ed classes. I volunteered to pass on my limited skills to the students. This resulted in my revised version: "Those who teach, learn."

Meat and Potatoes

There are a lot of miscellaneous machines and tools at Grant-Kohrs which were not original to the ranch. The park service acquired them to use, either for actual work or as "props" for interpretive programs.

It was a great advantage to work in the ranch blacksmith shop in the nineteen-eighties and nineties. We had many visitors who, in the 20's and 30's, had worked with machinery just like our aging relics. From them, I began to learn the names and uses of many of the items, and they always seemed to come with a story.

One was a potato planter. It was filthy, clogged up with mud, scraps of wire and twine, and the canvas bags which had once held the seed potatoes were in tatters. A small plug of wood had been whittled out to replace the missing grease fitting on one wheel. It seemed to be just another broken remnant of agriculture's past in the Deer Lodge Valley.

However hard the work might have been when these ranch visitors had actually been using such a planter, they always smiled when they saw it. I learned in time they weren't really seeing our rusty wreck. They were seeing the team of horses that pulled their planter. They were seeing their father or grandfather driving the team. They were seeing themselves as youngsters, walking along behind, stomping on dirt clods which had not been sufficiently broken up by the disk coverers at the rear.

"Remember?" they would ask each other. "Remember that team we had….what were their names?""Oh yes! Susie and May!" Each recollection led to another one and I'd shamelessly eavesdrop, realizing that a rusty, broken-down planter had given them back a bit of their past.

Often, it would also connect them to the future. A wide-eyed grandchild would exclaim, "You actually used one of these?!" I could imagine these children, grown up and bringing their own children to the ranch. "You know," they would say, "your great-grandpa used one of these planters." "Really?!"

Their memories made me curious. I decided to do some repair work on the planter and moved it outside the blacksmith shop. The first and easiest thing was to replace the torn and rotten bags. They didn't have to be fancy. They had obviously been replaced before, possibly many times. The bags were clearly not store-bought. Someone had seemingly laced them together by poking cotton twine through the canvas, perhaps with a nail. Who had replaced them? When? Where? I supposed it was probably the same hired hands who had cut up the seed potatoes, planted and finally harvested the crop at whatever ranch the planter came from.

Though we don't have an original planter, we know potatoes were grown at the ranch. From small plots to a 25-acre field on the west side of the river, potatoes were a ranch staple and another source of income.

Scraping off the mud and pulling wire and twine out of the mechanical picker wheels was the next job on my list.

Each repair brought more memories and advice from watching old-timers and more questions from visitors. One day I got the reels to actually turn. Vigorous cleaning with a wire brush freed weights which were designed to knock the seed potatoes off the pickers. The seed pieces fell down through the furrow opener and were covered by dirt tilled up by two disks at the rear. The repairs reached the point where it could actually have been used, and I had high hopes, but it didn't really fit in with our theme of the open range cattle era.

RESTORING THE POTATO PLANTER WAS AN EDUCATION IN ITSELF.

Or did it?

The answer is: It most certainly did.

This was meat and potatoes country. That did not come about simply because people had a taste for meat and potatoes. It was the result of the soil, the weather, the water, the length of the growing season, the native grasses, the sheltering hills and the northern European heritage of many early settlers. As early as 1916, the value of Montana's potato crop was nearly $6,000,000. That would equal about $132,000,000 today.

There is an overwhelming list of connections to that one broken-down piece of machinery. It could be used to compare the planting of root crops with tribal harvesting of bitterroot on the west side of the ranch. There were family farms and ranches which produced just enough crops to survive and there were thousands of acres of seed potatoes grown commercially in the Deer Lodge Valley in the 1930's and 40's.

I tracked down one former employee who had planted and harvested potatoes from a 25-acre field, and hearing his stories gave more dimension to all the other hands.

A nearly unbroken pay record exists of the hired hands who worked at the ranch, but with few exceptions, little is known about their specific chores. "Ranch work." That's about all the records say, but anyone who has worked on a ranch knows it includes virtually every sort of labor, from irrigating to haying to branding, calving, blacksmithing, feeding chickens, chopping firewood and more.

The challenge is to learn the stories, understand the objects, preserve them in their historic context and give them to future generations.

Research

The only thing wrong with the college I attended so briefly was that I didn't want to be there and they didn't want me there. It was a no-fault dropout situation. Education is a wonderful thing, but it doesn't all happen in school.

One amazing advantage of working in a new Park Service area is the lack of the dreaded Black Binder. In various guises, this three-ring bit of tyranny is handed to new employees with an injunction to read it and an implication that it contains all that is good and true about that particular national park. It's efficient and stifling.

So many years have passed that I can't recall why I first went to the Montana Historical Society in Helena. It was definitely out of character. I have a faint recollection it was because our files contained an 1886 newspaper interview of Johnny Grant in which he said he had 21 children and they had seven mothers. There was no Black Binder to turn to. The park knew of only three children and two mothers. My curiosity was not academic or professional in any way. I simply wanted to know if it was true.

Just inside the library door at the Historical Society was a man sitting at a desk. I couldn't have known it, but I was about to meet someone who would turn idle curiosity into a consuming passion for historical research.

Dave Walter was the research librarian. He led me through the intricacies of archival research, and when my non-academic mind failed to learn parts of the system, he led me through them again. It is tempting to say he never met a history question he didn't like. Certainly, he treated every question which came across his desk with interest and curiosity and respect. He gave people, including myself, the impression that we had just brought in an important and fascinating question, even if it was something as nebulous as, "My grandfather worked out West in the 1880s but we don't know his real name. Do you have anything about him?"

DAVE WALTER, MONTANA HISTORICAL SOCIETY: THE BEST.

If that initial visit had been less welcoming, I doubt I would have returned. Up until that point, the world of historical research had seemed as sterile and unappealing as a recently disinfected hospital ward. History, after all, was just a long series of wars, punctuated by inventions and injustices.

In the coming years I sometimes wondered if I had become a researcher who happened to work at the ranch, or a ranch employee who happened to do a lot of research. The entire staff at the Society was happy, interested, helpful, and had a deep love of history that was inspiring. My forays into the archives became so frequent I'm surprised I didn't wear a path between the ranch and Helena.

As research took me farther and farther afield, into archives from Canada to California to Washington D.C., I was also to learn that not all historical societies are created equal, but my mother taught me, "If you can't say something nice, don't say anything." So I won't.

There may be historians who are unaffected by personal bias and who can instinctively tell truth from fiction. I've never met one.

I did have one advantage in my research that kept me from unquestioningly seizing upon any stray anecdote which came my way. Law enforcement emphasizes "probable cause" and "reasonable suspicion." As a law enforcement ranger, I had been trained at the Federal Law Enforcement Training Center (FLETC) in Georgia and it seemed natural to apply the same principles to historic research as to a criminal investigation. If I had probable cause to believe the material I'd found was true, I made it generally available. If I only had a reasonable suspicion, I usually tucked it into a file until I could find corroborating evidence. Patently false stories also served a purpose, and were discredited but recorded to use as a reminder that just because a story was old didn't mean it was true.

Piles of information confront the enthusiastic researcher and the temptation to take it all as truth is strong. This is a bad idea. One of the deadliest excuses I hear for telling an unverified story is, "The people like it." Imagine conducting law enforcement that way: "Well, we don't actually have any evidence that the suspect committed the crime, but it makes a good story…"

Even with the most scrupulous care, truth is hard to come by. I once amused myself by imagining how convenient it would be if a tiny bell would ring whenever I stumbled upon an historical "fact" which wasn't strictly true. Then I imagined going mad at the constant ringing.

Con Warren: An Introduction

Conrad Kohrs Warren will have to come into this story in phases, because that's how he came into my life.

Perhaps I should say, that's how he let me into his life. The establishment of Grant-Kohrs Ranch was a very public event, but he was a very private man. I met him first as the Great Man, the man who had the vision to preserve his grandparent's ranch; the history, the buildings, the land and the artifacts.

The initial Park Service purchase was a little over 200 acres and included all the key historic structures and thousands of artifacts and documents. The historic ranch was like the hole in a donut and Con continued to own and work the land around us, which included his home and the buildings he had put up since the early 1930s.

CONRAD KOHRS WARREN.

Oil and vinegar may make a good salad dressing, but they don't mix. There is a little of that—sometimes a lot of that—between ranchers and the government. And just imagine: Now his ranch was part of the government, subject to rules which didn't always make sense from a cattleman's viewpoint. Furthermore, there were times when we acted as if he had come with the ranch. Our questions were ceaseless and tended to ignore the fact that he hadn't been around since the 1860s.

After a time, he became very reluctant to be interviewed, feeling that we expected his every pronouncement to be for the ages.

I had not yet started doing research and found few occasions to speak with him, but he taught me one vital lesson even before I got over my initial "cattle ranching is all about horses" phase. Seizing any excuse to ride, I was under the impression I was helping him move some of his Hereford cattle through the historic ranch to his pastures on the west side of the Clark Fork River. This involved crossing a bridge and the cattle balked.

Trying to be a Real Cowgirl, I whooped and waved my arms around and pushed my horse at them.

"No," he said, quietly. "All you have to do is make it hard for them to go the wrong way, and easy to go the right way," (and here he spoke calmly and deliberately, probably using the same tone of voice he'd use on a cow) "and give them time to make up their minds."

Such a simple statement and so important at the same time.

If you're trying to move cows down the alley from Corral A to Corral Z, it will go a lot easier if you close the gates into Corrals B through Y first! Otherwise, you will be losing cattle through every wrong gate.

And giving them time to make up their minds is just as critical. Standing quietly in the direction you don't want them to go is usually more effective than trying to hurry them in the right direction.

Con was a Great Man. It was a long time before I began to know and understand him and to realize that his worth had little to do with the mission of the Park Service and everything to do with his heart and mind. That story will unfold later. After all, process that took years can't be tucked into one short chapter.

About those Cows

Since Con introduced the subject, this is as good a time as any to point out that cattle do have minds. It used to irritate me no end that the cows could be up to their ears in good pasture but as soon as they had a belly full, they'd break out. But grazing animals understood better than I that pasture doesn't last forever. Their ancestors figured out long ago that moving around to graze was to their advantage. They didn't graze the grass down to the dirt. They didn't drink the water hole dry. They didn't become a stationary feast for every passing predator.

"Stupid cows!" we'd growl. "They couldn't even find the gate!"

Cows don't build gates.

"Dumb cow! I was just trying to doctor her calf and she nearly ran me over!" Well now, how is a cow supposed to understand antibiotics? Besides, if she's had calves before, she knows that some of that "doctoring" makes her calf bawl with either fear or pain. Sometimes the humans are there to snatch her calf away from her forever. And if cows were really smart, as humans judge it, they'd probably trample us before we could eat them.

One of the smartest cow maneuvers I ever saw occurred at the start of calving season. Only one cow had calved, and the whole herd was at the far end of the field south of my home. Something was going on out there. It looked like mob action. Through my field glasses, I saw that a full-grown steer had broken into that pasture and was trying to harass the calf—not unusual behavior.

Nearly all the cows were facing him down, the mother in the lead and nearly nose to nose with him. Meanwhile, another cow was quietly leading the tiny calf into the brush. Only when the mother was certain her offspring had been safely stashed out of sight did she end the confrontation. The mother's fight, the herd's backup and the sly removal of the calf impressed me. Some people will never be convinced that cows think. They'll insist it's just instinct. Maybe we could compromise: Call it instinctive thinking.

In any event, their handling of the crisis was wiser than mine in a similar circumstance. I heard a lot of bellowing and discovered that a cow had calved so close to the fence that her calf had shot under the fence rails and was in with some four-year-old Longhorn steers. They were sliding their horns under the sopping wet newborn and flipping it into the air. I grabbed a stick, charged into the corral, whacked the nearest steer on the nose with it, wrestled the calf under the fence rail and back to its frantic mother.... and realized that the cottonwood twig I'd grabbed had been as big as a pencil and had broken in half when I'd whacked the steer. I'm sure half a ton of horned hostility had really been impressed. It simply ambled off, probably thinking, "Geeze, what a grouch!"

Generally speaking, cattle aren't vicious. Some can be as gentle as pets. Some can be gentle for years and suddenly become a hazard. Half a ton to a ton of beef can be intimidating. I kept expecting to be stepped on, but somehow, they never did. Early one spring, shortly before calving began, I finally figured out why. I was looking at a pregnant cow, head on. She was comically wide. She looked like a giant lollipop on two sticks. I realized that unlike horses, whose legs come straight down from the shoulder, cattle bulge out on the sides, their legs inset far enough that they might bump you with a shoulder or a flank if they pass by closely but their feet will miss you unless they decide to kick en passant.

Probably the most important thing I came to understand about cow temperament (and there are a lot of variations) is that whether they're hand-raised or part of the herd, unafraid is not the same as friendly.

There are friendly cows. Sweetpea, a registered Longhorn, would groan pathetically and drag herself up every night during calving when I came to check the cows. She'd walk along behind me, grumbling the whole time, but I appreciated her company. As long as she was with me, the other cows assumed I was just an ungainly member of the herd. Other cows have provided backup over the years—Connie, What-A-Butter, Penny Rose and I-5 among others.

There are also unfriendly cows. I named one bottle calf "Nandi," which was purportedly Hindi for "The Joyful One." A less joyful calf would be hard to imagine. She refused to be halter-broken and if she could talk, I imagine she would have said something along the line of, "I'm a cow. You're not. Just give me the darned bottle and go away!"

And then there was #219. I have a plastic model of a demented-looking cow. It has a numbered tag taped to its ear: #219. I bought and tagged it in memory of a cow which was courteous enough to let me live.

The epic had begun the winter before. I had the early morning calving check. This involves walking through the herd and checking for signs of impending motherhood. There's more to it, but let it suffice for now that as I stepped into the corral, #219 stretched her head up like a giraffe. This is not a good sign, and I gave her a wide berth as she had a brand-

new calf at her side. She had been part of the herd for several years, and had never been hostile before, but being "high-headed" is a warning. I left a note for the next person who was scheduled to check. "Watch out for #219!" This individual, with many more years' experience calving than I had, was inclined to disregard the warning—until she ran him down—a fact he didn't confess for several years.

It's a good idea to sell dangerous animals, but the observation, "she's always had a good calf" causes misbehavior to be overlooked on many occasions. She settled down in a few days. The next winter I started preparing for calving season by trying to make friends with her. Every time I was in her territory, I'd toss her an armload of hay. "It's just me, #219," I'd say. "You know me…"

Calving season arrived, and I came out in the early morning dark to check the herd. I went straight to the pens without checking the notes in the office. There was #219 with a brand-new calf. I would have avoided her if possible, but there was a cow behind her which needed looking at. I grabbed an armload of hay and strolled up slowly and obliquely. "It's just me…" I began the litany.

She watched, but didn't get up. I eased around her, checked the other cow, went back to the office and was met with a LARGE sign saying, "Don't go in with #219. She tried to kill two people today!!!!!"

Later that day I heard a bellowing and a tremendous crash as she tried to break through a solid board fence to get at me. We sold her immediately, but I've always been grateful that she gave me that one "home free" chance, a pretty fair return on the investment of a few armloads of hay.

Cry Fowl

It came as something of a surprise to learn that the Kohrs' dinner on very special occasions was likely to be turkey. It didn't seem right somehow. Shouldn't a famous cattle baron serve a "baron of beef" to his company?

His granddaughter, Anna Warren Bache, recalled, "At 1:00 p.m. on Christmas Day, dinner was served. The turkey was stuffed with apple-raisin dressing and served with gravy and lingonberry and cranberry sauces."

We had Rhode Island Red chickens when I arrived. They were considered to be an appropriate breed for the Kohrs era. The hen yard was big, with two coops. One was for the hens and roosters; the other was a brooder house for chicks.

I'd tried raising a couple of chicks when I was a youngster, but as soon as they reached the gangly, "teenaged" phase, they lost their appeal and I gave them away. With so much to learn about cattle and horses, I was somewhat taken aback to learn a person could spend years learning about something which seemed as paltry as poultry.

Though we at the ranch called it the "hen yard," we have also raised other birds from time to time; turkeys, ducks and geese. There was even a Guinea fowl that periodically whirred through the air like a feathered cannonball. Turkeys are fine. Ducks are funny. Geese should only be raised by deaf martial artists.

Roosters may or may not be mean and the same goes for turkeys. Ducks generally have too good a sense of humor to be unkind, but geese would be born with chips on their shoulders, if they had discernible shoulders. Furthermore, they can—and do—make a deafening racket. A neighbor living fully one mile away could hear our geese at night. As the only human living on site, I found them extremely annoying. When I complained, folks said, "Oh, but they make great watchdogs." This is true. If you want to know every time a bird flies overhead, each time a fly buzzes on the windowsill and you have no objection to a deafening burst of honks whenever a leaf blows through the yard, get geese.

GUINEA FOWL ATOP THE HENHOUSE.

IS THE COFFEE READY?

As for the chickens, most were just members of the flock, but gradually distinct personalities emerged. Miss Kitty, Mr. Mom and Sir Galahad were standouts.

In the early years after the ranch became a park, the hens could wander around outside their fenced yard during the day, and obligingly return to the coop on their own as evening came—except for Miss Kitty. She insisted on being walked home. If I didn't come out at the proper time to escort her, she'd arrive on my doorstep, clucking and scolding. I'd grouse (no relation) as I'd walk her back, saying, "You know perfectly well where the gate is. Why don't you just go home on your own?!"

She was wiser than I knew. One evening I walked her in the henhouse door and saw a skunk across the room, crouched under the nesting boxes. Skunks eat chickens. The other hens were already up on their roosting poles and dusk was settling in. Once it's dark, hens become sitting ducks and they go completely still when they're held upside-down. I grabbed two by their feet and after an outraged squawk, they dangled silently, heads down, as I carried them to the brooder house, came back and grabbed a couple more. I'd made several trips and was standing in mid-yard when the skunk decided to make a break for it. It darted past, perilously close to the hens hanging helplessly from my hands. As soon as it was out the gate, I turned around and returned all the hens to their own roosts.

I can't recall how Miss Kitty got her name, although she did come when we called, "Here, kitty, kitty."

Mr. Mom was a banty rooster. There are several breeds of bantys (or bantams, if you prefer). They're small, but they pack a lot of character. Mr. Mom got his nickname from the fact that he would take over the egg-setting duties of hens who abandoned their nests.

Sir Galahad was a banty too, a white silky. When he'd find a particularly tasty treat, be it food scrap, bug or mouse (chickens eat anything) he'd call his harem over to enjoy it. That was somewhat typical rooster behavior, but while most roosters were just using the treat as a ploy to lure the unsuspecting females within range of their own lascivious intentions, Sir Galahad stood by politely while the hens ate.

Although I counted the birds every evening when I tucked them in, they shifted around so busily that I sometimes missed a hen, but Sir Galahad never did. He'd stand outside the henhouse door until all the hens were inside, and if he refused to come in, I knew to search the yard for an elusive member of the flock. One time, I knew my count was accurate but he remained outside. I stepped to the door and found

HENHOUSE.

him facing down four barn cats, the largest of whom had been under suspicion for stealing chickens.

They weren't all perfect. There was a rooster I nicknamed Rocky whose attacks were more irritating than harmful. I finally took to placing a plastic bucket over his head when I'd enter to the coop. Then I'd fill the feeders and waterers in peace while an animated plastic bucket careened around the coop.

Barn cats don't typically bother the domestic fowl. The chicks and ducklings are usually well-guarded by their mothers, but even without their vigilance, there seems to be an understanding that ranch poultry are not legitimate cat-prey. We did have a big, gray tomcat who was caught several times eating eggs, deep yellow yolks dripping off his whiskers. He had a nice glossy coat, though.

It's hard to know where to start writing about chickens, and when to stop. Horses arrive with some ceremony. They have to be named or renamed. A horse which has been hard to catch, tends to bite and kick or tries to scrape a rider off on a fencepost may come with a socially unacceptable name. Cattle have precise seasons for calving, grazing, winter feeding, being branded and so on. But chickens…. Even though there's the delivery (by mail, in a cardboard box) of just-hatched, peeping chicks and other "events," such as molting and egg laying, chickens are sort of a continuum. I like them, but except for the occasional standout, they are generally a simple fact of ranch life, like the return of redwing blackbirds in late winter or the spicy smell of cottonwood leaves in the fall.

Of course, that's only one person's opinion. There are chicken fanciers who would argue the point, and those who would risk their lives for a hen. Just ask Sir Galahad.

Hay

What is there to say about hay? Just dried grass that you feed to cattle and horses, right?

My introduction to hay when I came to the ranch involved little more than feeding it to the livestock. To say this gave me an understanding of hay would be like saying that eating a slice of cake could give me an understanding of baking.

A short list of critical hay knowledge would include growing, fertilizing (or not fertilizing), irrigating, cutting, curing, baling or stacking, storing, feeding, grazing, harrowing and starting over the next spring. That's just a start! Whole chapters could be written on how to maintain your balance on the back of a truck which is bouncing across a field while you slice open bale strings and toss out flakes (natural divisions within the bale) without stabbing yourself with the bale splitter, without dropping any strings to entangle the following cattle, without being impaled when a Longhorn steer with three feet of horn on each side decides to hook his own bale off the truck…it's endless.

I have two bale splitters. One is a classic: a sharp, triangular sickle section off a mowing machine, attached at one end of a handle. At the other end is a long, sharp hook. Efficient as it is, I longed for something which was unlikely to slice me and which—if it fell off the back of the truck—wouldn't go through a tire or an animal's hoof.

Bob Hitt, a Helena knife maker, came up with a beauty. It looks like a large tuning fork. It's a perfect example of work by someone who could use both his head and his hands to solve a problem.

In the absence of either of those tools, a pocketknife —— easily lost and quickly dulled by cutting plastic twine—could be pressed into service. A length of twine poked under the bale string and sawed back and forth would generally saw through fairly quickly.

That's just string-tied bales. We still had wire-tied bales up Gold Creek, and losing a wire (or even worse, a piece of that wire) could be deadly if a cow ate it.

Until I began to have some responsibility for the care of animals which depended on hay for their existence, it was easy to think no farther than whether there was hay or not.

As with every other aspect of ranch work, I found there was always more to learn. There was, in fact, so much to learn that it's hard to know where to start.

Perhaps the best place to begin is in a corner of the bull pasture. Cows and bulls aren't kept together all year long, but that's another story. Suffice it to say that the bulls and steers are generally kept in a low pasture northwest of the ranch buildings.

There's a tiny flower that grows there every spring. It's called blue-eyed grass, although it isn't actually a grass. It is rather inconspicuous with a beautiful little blue flower and long slender leaves. It is soon eclipsed by the various grasses which spring up around it.

Knowing where I would find the blue-eyed grass, the bitterroot, Rocky Mountain iris and death camas was something that came to me gradually over many years. Noxious weeds such as knapweed were easily spotted, but insidious invaders such as leafy spurge didn't appear until later.

And what do these plants have to do with hay?

Well, bitterroot, which is the state flower, tells us by its very presence that it is in sandy, well-drained soil. The area won't receive a lot of natural moisture, either as rain or a high-water table. It won't be a good place to grow hay. Knapweed is a virtually inedible invasive weed which moves into ground which has been disturbed by injudicious plowing, overgrazing or unnecessary vehicle use. It is generally found on ground which is not receiving enough moisture to allow grass to compete with it and it can have a toxic effect on adjacent plants. Death camas (no subtlety there!) is extremely toxic and Rocky Mountain iris can be as well. Livestock avoid them unless no better feed is available. Overgrazing an area with such plants in the mix would put our stock at risk.

Each plant in the hayfield which is not a grass tells us something about the quality of the hay. It tells if the irrigation system is efficient, if the soil is lacking in nutrients, if the field is grazed the right amount at the right time.

And what about that blue-eyed grass? It became a friend to look for each spring; just a tiny part of hayfields which cover over 700 acres, providing grazing for livestock, hay for winter feed, soil and water for native plants and cover for wildlife.

As the old saying goes, "And that ain't hay!"

Weather

A long-ago employer sent me to Los Angeles to take a management course in a fancy hotel. Under one roof were my hotel room, the classroom, restaurant and several shops. I had polished my shoes on Monday, and at the end of the week realized they hadn't needed to be polished since. There wasn't a speck of dirt on them. It wasn't too long after that I found more congenial work—work which led outdoors.

When you work outdoors, you notice the weather. When you work in agriculture you notice it with anxiety or hope or frustration or joy. The weather isn't just something which affects the shine on your shoes. It affects your whole world. How deep is the snow pack? Will there be water for irrigating? How many cows will the pasture support if it doesn't rain, and if cows need to be sold, when? How many? Which ones? Will there be a decent market or will so many other ranchers have to sell that the price will bottom out? Will there be wildfires in the mountains? Will summer range burn? Is there going to be subzero weather at calving time? Will an early frost hurt the potato crop?

The weather matters every day, all year. The Deer Lodge valley would be a desert if it weren't for snowmelt from the surrounding mountains. There's a spot on Mount Powell which is watched with some anxiety in the summer. If that particular cleft still has a patch of snow on the 4th of July, enough irrigating water will get us through the summer. The weather is, in some ways, a partner. In a good year, it is a partner which receives due credit for its contributions to a good hay crop, healthy livestock and low fire danger.

There's practically a ritual associated with watching and discussing the weather. There are two distinct ways of interpreting it. If we don't get snow early in the winter, some folks start predicting drought, while others will say reassuringly, "Relax. We get the important snowpack in February and March." If we get a lot of snow, some people predict flooding and others say, "Don't worry. If it warms up slowly the runoff won't be so bad." There's always room for optimism. If we don't get the late winter snows, we can still get plenty of spring rain. For some folks there is always room for pessimism: If it rains when we've got hay cut, it can rot in the field.

Sometimes it can get pretty grim. One year we hadn't received the snows, and the rains hadn't started. It was time for the grasses to start their annual growth, and they did, but not with any vigor. If this

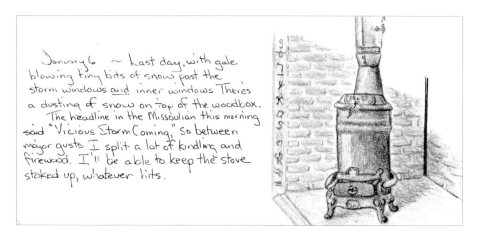

January 6 ~ Last day, with gale
blowing tiny bits of snow past the
storm windows and inner windows. There's
a dusting of snow on top of the woodbox.
  The headline in the Missoulian this morning
said "Vicious Storm Coming," so between
major gusts I split a lot of kindling and
firewood. I'll be able to keep the stove
stoked up, whatever hits.

continued, the grasses would head out (make seed heads) and stop growing. The possibility of a short hay crop could be read in grim faces and silence, as if talking about the situation could make it worse. Day after day was cloudless and dry. Then, one afternoon in late May, the rain started. All up and down the valley, people phoned each other to say it was raining. Of course, we all KNEW it was raining, but that wasn't the point. We wanted to say it out loud. We wanted to share the news. We wanted to celebrate.

That was when I realized one of the fundamental differences between urban and rural life. Having your comfort depend on the office air-conditioning means nothing compared to having your home, your family, your crops, your livestock, even your future depend on the weather.

Con Warren told me once that it snowed 23 inches on May 23, 1938, and the moisture "broke the back of the drought of the Depression." Half a century after the event, he could remember that snowstorm. I imagine there were more than phone calls up and down the valley that time. I imagine there were tears of relief.

Not Quite Self-Sufficient

When my government home of 27 years, an unprepossessing single-wide trailer, was unceremoniously hauled away, it showed a few signs of wear and tear and one purple stain which probably perplexed its next owner. It even perplexed me.

It was on the ceiling above the stove. Examination through a hand lens revealed that it was probably a blob of huckleberry juice. I wouldn't have been surprised had one of my early jelly-making ventures resulted in exploding juice, but a single blob was puzzling. It was in the best family tradition, however, as my mother had once decorated her kitchen ceiling with asparagus shot out through the vent of a pressure cooker.

I don't cook. I eat. However, in my attempt to understand and become part of the agricultural community, I had taken to picking wild huckleberries and making jelly. I was too old for 4H, so entering jelly in the open class at the Tri-County Fair seemed like a proper initiation into the rites of late summer.

Contrary to my childhood memories of Telegraph Creek, huckleberries do not actually smell like skunks. There were a few huckleberry bushes in the woods surrounding the cabin. These were what I would now call "snackleberries." That is, there were enough to munch on as you wandered around, but not enough to pick for a pie or jelly making. There was always a faint skunk smell in the same territory, and I came to associate the berries with the skunks.

A huckleberry is like a blueberry, but smaller, less abundant, and on steroids. Huckleberries hold pride of place in Montana hearts, though there are lots of other wild fruits to choose from: thimbleberries, chokecherries, tiny, sweet wild strawberries, serviceberries, elderberries, Oregon grape—it's a long list.

It's an admirable ambition to grow your own food and preserve it. Picking wild berries does not actually fit into that category. I did not, at first, appreciate the difference. When folks lived off the land, there may have been considerable satisfaction in putting by enough food to get them through the winter and tough times, but it wasn't an option. If a crop failed, they couldn't simply say, "Oh well, we'll pick up a pizza on the way home."

Even berry picking, while possibly enjoyable, had an element of necessity to it in the old days. These days it means spending the day in the mountains, searching for a good patch, taking along a lunch, enjoying the views and letting the dogs splash around in the creek.

Of course, there can be adventures. Every snapping twig could herald the arrival of a bear to contest your right to the fruit. I only encountered a bear once. It was a young one and we simultaneously popped up on opposite sides of an unusually tall bush. I never found out if it dropped its berry bucket as it ran off, but I know I did.

My biggest mistake picking huckleberries was in showing the dogs where the berries were. They had been constantly underfoot, begging, and I said, "Look! Pick your own berries!" It took them only a moment to grasp the concept—along with nearly every berry below knee height.

Conrad Kohrs would annually send two men up into the hills to pick huckleberries. This was not for any such frivolous purpose as jelly making. It was to provide his hired hands with food during a roundup or trail drive when they weren't able to get back to the chuckwagon.

Every year in late summer, a couple of fat steers would be butchered and the meat smoked over a willow fire. A few strips of this jerky, a handful of huckleberries and a dash of salt would be put into a little canvas bag. The bag would then be hit with a mallet to smash the berries and coat the meat with its juice. According to Con Warren, Kohrs employees made hundreds of these sacks. At the tender age of twelve, he was sent, alone, to herd some horses from the home ranch up to Augusta, about 100 miles away. All he had to eat on the journey was this "pemmican," as he called it. He told me it tasted all right, but "you chewed it a good long while." True pemmican, as the native tribes prepared it, was quite different from jerky. The meat in pemmican is pulverized and mixed with fat as well as various fruits to make it palatable. Jerky, in my opinion, is excellent for patching holes in leather boots.

In any event, entering jelly in the fair fell short of my mental picture of living off the land, but participating in the fair had its own rewards, acquainting me with members of the farm and ranch community who only tended to socialize in town a couple of times a year. Still, my own "crops" (wild fruit and garden flowers) missed the mark. It was a long time before I realized that home-grown food would provide nourishment which store-bought food did not—no matter the nutrition content.

A Lucky Loss of the Toss

I was never particularly ambitious. Since I was 12, I had wanted to be a ranger, and now I was. However, I was brought up with the Great American Myth: Anyone can grow up to be President.

Something which was never even discussed was whether anyone should want to be President. Still, success was often held to be synonymous with upward mobility.

When a supervisory job opened up at the ranch, I applied for it, motivated by many bad reasons.

There comes a time when the unthinkable happens: Your boss is actually younger than you. Dreadful. You are older. You should be the boss! This is nothing but ego.

Of course, a higher rank pays more, but I was blessed with a mother who raised us to believe that the Great Depression was not necessarily a one-time event. No matter how little my brother, sister and I earned when we went out on our own, we always saved some of it. And if we could manage to save on a small wage, did we really need a bigger one? This is not to say that we weren't happy to move up the economic scale.

My only theoretically good reason for wanting the promotion would have been the worst reason in the long run. A higher-ranking ranger should be more of a ranger. Right?

Wrong. When the Chief Ranger informed me that it had gotten down to the flip of a coin between me and another candidate and my rival "won the toss," I did not immediately realize that she had just done me the greatest favor of my career.

Generally speaking, a supervisory ranger is not more of a ranger, but less. Less time in the field. Less time with the visitors. Less time to become immersed in the park until you are nearly an organic part of it. You become more of a supervisor. More of an office person. There are exceptions, of course. There are rangers who moved up the ranks and brought out the best in their employees without losing what was the best in them. I have been fortunate enough to work for outstanding interpreters: Dave Karraker, who could call a wild bird to his shoulder with a simple whistle; Jim Warner who seemed to know everything but taught me to get the answers myself; Matt Conner, who could probably interpret to a lump of coal and inspire it to become a diamond.

However, I would have been a really rotten supervisor. Missing the work I loved, I would undoubtedly have wanted all my employees to do things exactly the way I would, were I lucky enough to still be doing it.

Had I understood then what I understand now, I would have bought a two-headed coin for that toss—and I would have called "tails!" As it happens, I won anyway.

Having, as I then thought, "lost" the promotion, I gave myself two choices: Move on or find another way to grow. The ranch already had a tight hold on me, so I decided to try a little freelance writing. Three decades and more than 1600 newspaper columns later I still consider myself an aspiring writer.

It had never been my ambition to become a writer. It was Mrs. Schneider's ambition for me. She was my English teacher in junior high, and she forced me to write the class poem and recite it (with churning stomach and wobbly knees) in front of the entire student body.

It is often claimed that seeds found in ancient Egyptian tombs have germinated. Mrs. Schneider's prediction that I would become a writer didn't take thousands of years to germinate. It was more like the emergence of a 17-year locust.

The truth is, I got more out of my articles than the readers. A 500-word deadline every week meant I couldn't ignore the least thing that happened. Sometimes I had nothing but a single word to get me started—the name of a flower, the day's temperature or just "Ouch!" Without the newspaper deadline I could have done my park ranger job satisfactorily but remained oblivious of my surroundings. I would have missed a lot. Finding myself unable to think of anything to write about was a periodic wakeup call. It reminded me I was in danger of slipping into a routine. When really stuck, I'd go out and do something just so I could write about it. Taking a walk on the "wild" side of the ranch, communing with the barn cats or working in the garden always led to some new observation. Forays into Montana newspapers of the 1800s were interesting and added to my large collection of chicken jokes.

Making up a story for the column was not an option. It's hard enough to remember the truth. Lying would have required a better memory that I ever had.

Community

When a stranger comes into a small town, people notice. It's almost a game: Where do they come from? Are they passing through or are they settling here? Are they in agriculture? If they're dressed "western" are they horse or cow people? Engaged in this activity, we sometimes forget that the strangers are looking at us too, and our speculative gazes can be unnerving.

My friend, Matt, was noticeably tall and new to town. He commented on what seemed to be unfriendly looks. "They're just wondering who you are," I told him. "Next time, just smile and say hello." It worked.

I have heard people say, "I've been in this town 20 years and I'm still an outsider."

Perhaps people will always recall if someone is an "immigrant," but that doesn't mean they can't be part of the community—or "communities."

There are no engraved invitations to become part of a community, but there are plenty of opportunities. Many of them involve volunteering, and my coworkers found their skills and interests in great demand: Peggy led 4-H. Willy coached Little League and basketball. Keith conducted free housing inspections for seniors. Staff members were volunteer fire fighters, ambulance crew, chaperones at school dances and more. The community needed them and they stepped up.

My interest was centered on the agricultural sector of the community and particularly the Ag-Ed classes at Powell County High School. This narrow field of vision meant I didn't realize that the town wasn't here merely to provide a commercial hub for the surrounding ranchers, loggers and miners.

A friend who had a country band asked to borrow my banjo for a man who occasionally played with the band at local bars. He often asked me to come hear them play, and I finally did. The bar was crowded, not with n'er-do-wells, but with a young and happy crowd of people I'd never seen before, yet they obviously were local. They all knew and greeted each other. Where, I wondered, do these people hide during the day?

The answer was, "right here" —working in stores, at the hospital, the courthouse, the local schools, the lumber mill. There was an entire town full of businesses I rarely had occasion to enter. These people, though strangers to me, had jobs, families, went to high school football, basketball, volleyball and other games, bowled in a league and more.

They, too, were volunteers in the community. The local food bank, literacy programs, the community garden, senior citizen advocacy and historic preservation each had its cadre of volunteers.

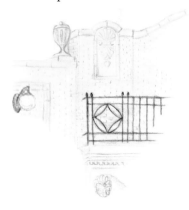

One of the most magnificent community efforts came about when the lovely Rialto Theater, built in 1921 and gutted by fire in 2006, was not abandoned as might have been expected. With an estimate of over $3 million dollars to restore it and a population of not much more than 3,000, Deer Lodge set out to raise the money and rebuild. They succeeded, winning grants, obtaining donations

AFTER A DISASTROUS FIRE, THE COMMUNITY WORKED
TOGETHER TO RESTORE THE ICONIC RIALTO THEATER.

large and small and investing no little amount of sweat equity in the effort. I doubt that anyone working to clear the charred wreckage, clean and paint and rebuild ever asked the people working beside them if they were natives.

Natives may be born, but members of the community are made: self-made.

Respect for Labor

When the Park Service brought in an old house trailer for the first site resident, they told maintenance worker, Arnold Larsen, to bury the water line three feet deep.

Arnold, however, had spent many decades in the valley. He buried the lines six feet down, and in 28 years I only had frozen pipes once, and that time only for half an hour. That was the year it was 30 to 40 below for three weeks and substantial buildings in town were plagued with broken pipes and malfunctioning heating systems.

Good work, Arnold!

When I first started working in the blacksmith shop, the haying crew brought me a horse-drawn rake with a bent axle late one afternoon. They told me they'd need it by 9:00 the next morning, laughed and walked off, knowing I didn't have a clue how to fix it.

But Dave knew, so I phoned him. His advice was clear and precise. A few boards, a tall jack, a long chain and a sledge hammer were all I needed, and my gleeful satisfaction knew no bounds the next morning when they came in and said, "Well, have you got it fixed?" and I replied, "Yep!" and sauntered off.

Thanks, Dave.

Mike taught me about "holidays." That's what he called gaps in an otherwise smooth paint job when you leave a thin spot because you stopped for one reason or another and started again, not blending the areas together.

Fine advice, Mike.

Over the years, I began to harbor a pet peeve. (People think I only have pet cats, but in fact, I have a number of pet peeves.)

Some people whose skills are vital when we need them are disregarded when we don't. "A broken pipe! It's 2:00 Sunday morning! Omigosh! We've got to get a plumber here right away!"

And the plumber responds with the knowledge and skills and tools to solve the crisis and considers it lucky if the customer doesn't object to the bill.

But that is now. What about "then?"

Admonitions to respect labor go back at least as far as the Book of Luke: "For the laborer is worthy of his hire." In the Qur'an, it is stated that "No job is to be held in contempt or considered inferior. Honour belongs to those who work and not to those who sit idle and survive as parasites."

And yet, it is nearly futile to try to explain this to some folks. No examples and no philosophies are enough to open that narrow mental drawer.

The only solution is to ensure that everyone has the experience and satisfaction of working with their personal team, but…how?

The Dirty Hands Book

Among advantages most Montana kids have over youngsters in other parts of the country is their ability and willingness to get dirty.

Beginning sometime in the 1990s, we started to get young visitors from some of the more urbanized states who were actually unwilling to step into a pasture because they would get dirty! To their credit, it usually only took a few minutes before they overcame their aversion, but it was disturbing nonetheless. This need for cleanliness even began to invade Montana.

Yes, times change, and yes, it is possible (though not, in my opinion desirable) to live a nearly dirtless life, but how sad.

They miss out on that sense of discovery as they poke a stick into soft soil just to see what's underneath. They never turn over a rock and discover a new world of exposed roots, interesting bugs and even an occasional ant colony with ants carrying eggs to safety until the rock is gently replaced and the colony returns to normal.

This peculiar phenomenon began with school age kids, but they grew up, which added young adults to the sanitized lot. I finally realized that fully two generations of visitors had never pulled the center out of a stem of grass and nibbled on the sweetness of the soft inner core. They had never made an annoying whistle out of a blade of grass. They had never flicked a dandelion flower at a parent or sibling, decorating their "target" with its bright yellow pollen.

That's when I happened on the notebook.

It was in the supply closet, where I had gone for a pen. The cover was black, and my immediate thought was how handy it would be in the blacksmith shop, since it wouldn't show the coal dust.

But what could I use it for?

I'm not sure where the idea came from, but I started using it to keep a record of visitor answers to the question, "Why is work you do with dirty hands important?" By the end of the first day I had collected a lot of wisdom:

A xeriscaper who created cactus and rock gardens in the southwest replied, "Dirty hands save water."

A lively discussion was started by a visitor who asserted "People who get dirty don't get sick."

In a group of kindergarten and first graders one small child pointed out, "Dirty hands mean you can be a blacksmith and you can wash them later."

Then Tristen, an eighth grader from up in the Flathead north of Missoula waited until his classmates had left before confiding, "Dirty hands leave legacies." He dashed off to rejoin his class and left me stunned at his insight.

Over the next couple of years, I continued to ask the question, and visitors never let me down. A sampling of roughly two hundred quotes includes the following:

- Dirty hands are knowledge put to work. A NORTH CAROLINA COUPLE

- There is always work for your hands. VISITOR FROM THE NETHERLANDS

- Dirty hands mean you've been having fun. YOUNG VISITOR FROM NEW YORK

- My father said, "Never trust a man who doesn't have a little dirt under his fingernails." VISITOR FROM SOUTH DAKOTA

- Dirty hands mean your wife wouldn't let you take a nap. HUSBAND WHOSE WIFE HAD A SENSE OF HUMOR

- The neat thing about hands is they're covered with skin and they wash. CALIFORNIA VO-AG (VOCATIONAL AGRICULTURE) TEACHER

- Dirty hands feed the world. VISITOR FROM SOUTH CAROLINA

- If you ask me, if your hands are always clean, I think you're a boring person and have bad OCD. JOB CORPS TRAINEE

- The most important things that happen in the world; food, shelter, happen because of people who aren't afraid to get their hands dirty! CONTRIBUTED WITH VEHEMENCE BY A LINEMAN

One day I dropped the book and it flipped open as it landed. Out of the center fell a pressed flower; a dogtooth violet. It's a lovely yellow lily which blossoms at the receding edge of the snowline as spring returns

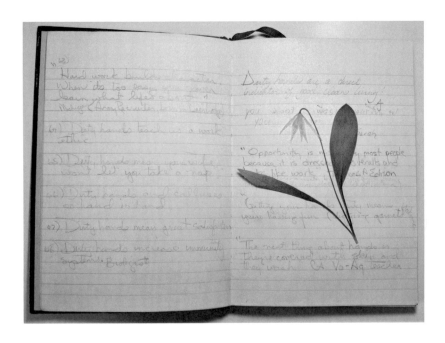

64

to the mountains. I had thought the book brand new when I took it out of the supply closet. Not a page had been written on, but as if in anticipation of its future use, someone had placed the flower there. Someone, I'm sure, who knew the joys of dirty hands.

Don't Ask if You Can Help

I'm guilty. I have watched, unthinking, as a coworker picked up something heavy and belatedly asked, "Do you want some help with that?"

I have seen my fellow employees walking back and forth, bringing in boxes of who-knows-what and only on the fourth or fifth trip has it occurred to me to say, "Is there more to come in?"

Those would have been steps in the right direction, but I still wouldn't have gone far enough.

Dawn enlightened me. She never asked. If I started to pick up something heavy, she was already lifting the other end. As soon as she saw me carrying something into the shop or the office she was checking to see if there was anything more to bring in.

Particularly at my desk, I tended to be oblivious of what other people were doing. I think a computer or a phone has an insidious way of becoming your coworker. You can be in the same room with people but you are in different worlds.

Dawn was raised on a ranch and was in charge of our ranching operation. She was a good role model, and even after she'd drifted off to Nevada, her lessons stayed with me.

That's the good side of the coin. The flip side is a tendency not to ask for help when you need it.

There are a couple of reasons for this. One is our own self-image as independent westerners. That image, however, grew out of the second reason: There is often no one around to help.

When you're over on the west side of the river, loading hay bales onto the back of the truck, you know very well that you aren't supposed to lift something "too heavy" by yourself, but that's hard to define. Furthermore, the cattle are waiting to be fed. It's the weekend and there is no one else around. You pick the bales up anyway.

Back in 1995, the government shut down for 21 days. As the only commissioned law enforcement person on the staff, I was the only employee considered "essential" for 17 of those days. This was not because the ranch was a hotbed of crime. It wasn't. However, in the unlikely event that there was a law enforcement problem, I was the only one who could respond.

My only truly "essential" duty was to feed the livestock. This meant loading 21 bales a day onto the truck. They were ridiculously heavy, and after a couple of weeks of this I dropped by the grain elevator in town and asked the owner if he'd weigh one for me.

As he reached for it I said, "Be careful. It's heavy!" Understandably, he gave me a look which said, "I've been doing this all my life. Don't worry." However, after one wide-eyed tug, he called one of his boys over and they carried it to the scale: One hundred thirty-eight pounds—about 10 pounds more than I weighed. He could, of course, have carried it by himself, but unlike me, he had too much sense to do so.

It was still necessary to feed the livestock each day, but now that my brain had been forced to accept what my arms, legs and back had been trying to tell me, I worked out a complicated and time-consuming system whereby I never lifted the entire weight of any bale. During a shutdown, furloughed employees are not even allowed to enter the property, let alone volunteer to help, but there were plenty of local folk who would have come out if I'd let them know the problem.

Once you get the hang of it, it's easier to know when to offer help. It takes a little longer to admit that there are times when you need to ask for it.

John Francis Grant

There's an old saying, "murder will out," which contends that in the end, evil deeds will reveal themselves.

On a much pleasanter note, I'd like to say, "History will out."

It was 14 years from the time I began my search for documentation of John Francis Grant's purported seven wives until the story of his life in Montana and Idaho was published by Washington State University Press as "Very Close to Trouble: The Johnny Grant Memoir."

I began by looking in the most obvious places. There were baptisms, census records and anecdotes by his contemporaries, but I got no farther than three or four wives and seven children.

What I had ignored from the first was the clue that in 1952, his grandson had sent several pages of Grant's memoir to Kohrs' grandson.

Twelve pages, in fact. They were part of our archives—a treasure house of ranching history spanning more than a century.

It may be that some long-ago school teacher tried to teach me the difference between primary and secondary sources, but I evidently wasn't listening. Even when I started working with primary documents, I kept trying to patch together Grant's story from church and courthouse records and the often-dubious recollections of his contemporaries.

At some point, it finally occurred to me that if there were 12 pages of excerpts, there might be more, but 1952 was long past. There was no address for the descendant and I only had Con's recollection that the man had lived in Canada. The trail was cold.

Extending the search to archives and museums turned up bits and pieces. An encouraging breakthrough occurred when I was in an archive in another state. They had once had a manuscript about Grant's father by a Grant relative, they said, but it had been missing since 1935.

As long as I was there, I decided to look through records of Grant's contemporaries, and in a box which evidently hadn't been opened since 1935, there was the missing manuscript. A long-gone researcher had evidently added it to his own papers which had been donated on his death to that archive.

Elated, I copied the pages, then returned it to the desk practically bubbling with enthusiasm over the "rediscovery" of the paper they had declared long-gone. This would have been a matter for celebration at the Montana Historical Society.

This archive was different. They shrugged, replaced it in the wrong box and sent it back down into storage. Oh well.

It was pure luck that I'd checked that box, but that sort of coincidence happens so frequently that I am no longer as surprised at such luck.

One of the advantages of the open and curious attitudes of staff and researchers at the Helena archive is that we tend to chat about our research topics, and—as often as not—someone will overhear and know something to the point.

The actual discovery of Johnny Grant's memoir (a 90,000-word treasure) was due to just such a coincidence, but the circumstances were so extraordinary that I have never yet written them down because I can scarcely believe them myself. How can I ask anyone else to do so?

The hand-written memoir had been carefully preserved by descendants and had spent several years in a safe deposit box in Canada.

The most ambitious researcher is unlikely to have access to the world's bank vaults. I can only conclude that the fates decreed the time had come for the story to be told.

I initially located only a typed transcription of the original, made I believe, by a nephew or great-nephew. As accurate as it proved to be, once the original was found, there were still subtle changes.

Johnny dictated the memoirs to Clothild Bruneau Grant, his last wife. She wrote in her foreword that while she would not change anything, she would leave some things out. Johnny, himself, admitted he would refrain from telling some tales which were, as he put it, "not just suited to the tastes of my fair readers." Tantalizing!

Perhaps I don't always know my limitations, but one thing was obvious; I did not have sufficient knowledge of Canadian history to edit his Canadian years. I only covered the Idaho and Montana years in the first published portions of his memoirs. It was left to a Canadian history professor, Gerhard Ens, to edit and publish the entire document.

Grant's memoir was the most extraordinary find, but it was by no means the only time the answer to one of our historical puzzles was found in an unlikely place.

We had debated for years whether the rough log section of bunkhouse row was Grant's home while his impressive trading post was being built. It was a bit unusual, consisting of three rooms in a row, with the center room open on one side for cooking. Our records couldn't settle the question. Then Kenneth Owens, a California history professor, discovered the journal of Edwin Purple in a New York City archive and Purple's account described that log house perfectly. Knowing my interest in Grant, he sent me the description in time to include it in Grant's memoir.

I'm often told I should do my research on the internet. It has its uses, but I only use it to help locate original materials. Using material from the internet itself is like panning for gold in a sandbox. The gold may be there, but it's somebody else's sandbox.

Calving Time

As mentioned earlier, it isn't enough to simply look at the cows as if they're just scenery. Especially at calving season, you have to really see them—from all sides. Gene—rancher, saddlemaker, backcountry packer, roper and rock-solid friend, helped me understand the annual ritual of calving.

At least in this neck of the woods, calving doesn't happen all year round. The bulls and cows are together for about 45-50 days, from about nine months before you want the cows to start calving; then the bulls are removed from the cow herd a bit more than nine months before you want the cows to stop calving.

The actual calving "season" does shift a bit though. When I started at the ranch, we calved (that is, of course, the cows calved) in January and February. That held true for most of the ranchers in the valley. There were two rationales for this.

- Calving early in the year meant you'd have bigger calves by sale time, traditionally in the fall. Since calves are sold by weight, you'd make more money.

- Frozen ground is less germy than wet ground. Calves are pretty tough when it comes to cold, so calving later meant wetter ground and more sick calves.

A couple of hard winters and high calf losses gradually shifted the calving season later.

The main point, whether it happens in late winter or early spring, is that bred cows calve. Sometimes they need help. Sometimes they don't. If you aren't watching, it doesn't take long to lose a calf, a cow or both if there's a problem.

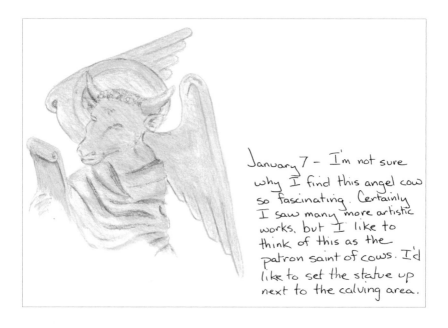

January 7 - I'm not sure why I find this angel cow so fascinating. Certainly I saw many more artistic works, but I like to think of this as the patron saint of cows. I'd like to set the statue up next to the calving area.

So, you watch. You watch at intervals 24 hours a day. The standard checks are usually every two or three hours, with particular attention to heifers (young cows having their first calves).

My favorite calving shifts were three a.m. or six a.m.

The only time I got a bit tired of the night calving shift was when the bulls had what you might call a field day for some time after they should have been separated from the cows. We calved for ninety-six days. That's a lot of early wakeups. On the other hand, it was nice as the weeks stretched towards spring and dawn came earlier and I could see better. Even a good-natured cow can get a little crabby if you trip over her in the dark.

What are the signs of imminent calving? Well, a cow lying down and having a calf is a pretty reliable sign, but there are plenty of clues before that.

"Bagging up" means the cow has started producing milk. This would be a more helpful sign, except an old Shorthorn cow may bag up days in advance of calving and a young Longhorn cow can be nursing a healthy calf with scarcely a sign of milk production.

Explaining all the signs of impending calving to non-agricultural visitors in a socially acceptable manner is a bit tricky. Fortunately, while there are physical descriptions to avoid in polite company, there are bovine behaviors which are inoffensive.

- If the herd shows up to be fed except one cow who stays down in the willows, watch.

- If the cow's tail sort of perks up at the base and the rest of the tail hangs down in the shape of an old-fashioned pump handle, watch.

- If she's restless, lying down, getting up, lying down again, watch.

- If she's sniffing the ground, watch.

- If she tries to steal another cow's calf, watch.

- If she's standing very still, her belly sort of pulled up and a particularly faraway look in her eyes, watch.

- If she's "wringing her tail" so it looks like a snake you have just run over with a bicycle, watch closely.

Those are just a few things to watch for, and you have to multiply it by the number of cows you have. Once you spot a sign of imminent calving, you stick around or increase your checks to make sure it goes well.

So, what if there is a problem? What kinds of problems are there? How can you tell and what can you do about it?

That's a chapter in itself, but I will just mention that if you are going to reach into a cow to straighten out her calf's leg, take your ring off first.

Birds Watching

Though I mean to be observant, I rarely spot any wild critters before they spot me.

Of course, they are often lurking in tall grass, thick brush, on top of trees and so on. I find it difficult to lurk as effectively. Tall grass is itchy, I sound like an enraged rhino when I try to slip through thick brush and bird-watching from the top of a tree would probably be a matter of getting a quick glimpse as I crashed down.

One bird down by the river didn't make an appearance until I'd been at the ranch for over 35 years. I knew what it was. Its song was very distinctive. It made me think of a tin whistle crossed with a calliope. A friend who knew all the local birds identified it by its song as a marsh wren, but years passed and I still hadn't seen one. I began to think of it as a disembodied song, lingering in habitat long-since given over to human use. One winter when the swamp was frozen over, I walked out to the cattails and found a marsh wren nest—a marvelous construction of woven cattail leaves, ball-shaped with a round opening about two-thirds of the way up its side.

I can barely fold a sheet evenly. How can a tiny bird build such an elegant nursery with just its beak?

Nearly four decades after I came to the ranch, I was walking by some willows and osiers and saw a flicker of movement. I could see it wasn't

January 2. What are the chances that I can average a sketch a day? This little wren was "cheating" - I took it from a sculpture I got years ago in Bath, England. It was a good day for birdwatching: Pairs of bald eagles soaring over the Little Blackfoot and lots of birds at the feeder, but they don't pose well.

71

the usual chickadee or sparrow. I stopped, and though expecting no response, said quietly, "What are you? Come on out. I won't hurt you."

To my amazement, it emerged, perching on a nearby branch, and giving me a few precious seconds to see it clearly, its tail nearly straight up in classic wren fashion. Then it dove back into the deep brush, perhaps never to be seen again.

Whether or not it reappears, it was a gift to see it once.

Sometimes I know birds are there because of the noise they make. Sometimes I know they're there because of their silence. There's nearly always a slight background of birdsong. If I step out to complete silence, it alerts me to the presence of a predator.

The mere presence of a human rarely quiets any but the closest birds. They are also quick to learn the common locations of resident cats, but occasionally the perceived threat will be one of my feline companions. The cats focus their attention (as they should) on rodents in the barns, but they like to practice their sneaking skills on bugs, each other, calves and other potential prey, including their accompanying human.

Some birds watch us warily, and some find us useful allies—or perhaps I should say "tools."

One evening I had driven the baler too quickly through some hay which had not dried sufficiently and the heavy, wet grass had plugged it up. This was early in my tractor-driving days and I was under the impression that simply stopping the PTO (the connection which allows the tractor's engine to run whatever machinery it's pulling) would make it safe to reach inside and clear the baler. I was wrong, but luckily I wasn't drawn in and baled.

What happened was that the roar of the tractor continued as I pulled clumps of hay out of the baler. When it was clear, I turned around and saw a sharp-shinned hawk sitting on a bale a few feet behind me, apparently hoping I'd also pull out a tasty mouse. Animals have an amazing ability to make use of humans. My presence and the noise of the tractor caused it no alarm.

When I drove off without proffering a snack, it took to the air, but kept a sharp eye on my progress in case I scared up anything tasty.

It wasn't the only bird which followed the baler. A mouse seemed to me to be pretty paltry prey to a bird with a wingspan of around seven feet, but a golden eagle often coasted along behind the clattering machine, borne up by the updraft of sun-warmed fields in the late afternoon.

Three members of the "corvids" inhabit the ranch: ravens, crows and magpies. With nary a magpie in sight, I can put a bit of dry food out for the cats which hang around the blacksmith shop, and half a dozen magpies will materialize in moments. They must post one magpie as a lookout. If I'm doling out a special treat to the shop cats, such as canned food a house cat has disdained, the birds hear the tap of the can and 15-20 of them will soar in from all directions.

Over on the west side of the river, I spotted something odd through my binoculars. It looked like an airborne tennis ball attached to the center of a small cannonball. On each side were short sticks which wagged up and down in unison. I finally realized it was coming closer and lowered my binoculars just in time to see a curlew shift its trajectory and soar right over my head instead of into my nose. I guess it just wanted a better look. When you're out birdwatching, remember: birds watch too.

A Change of Scene

There is a fascinating world out there. Though I have at least landed in about twenty countries and spent some time in a few of them, that's not much of a dent in a list of possible foreign adventures. When I'm in a foreign country I like to explore it without an itinerary. I walk around the great cities of Europe with no plans other than to head for anything interesting, historic, flowering or edible that comes into view. It's great fun.

However, that doesn't mean I'm bored at the ranch. If I want a change of scene, I can find it in just a few minutes.

Take, for example, a slice of hillside on the west side of the river. I won't even cheat and slip a section of nearby Mount Powell into my hypothetical slice for dramatic effect.

There are cottonwoods, birch and willow in the river bottom. As the hill slopes up there are irrigated pastures, irrigation ditches, a road, fences, dry hills and just a smidge of Montana's Pikes Peak showing at the crest. That's not my fault. It's hard to find a ranch view without some distant range showing.

Light accounts for a lot of the changes; bright morning light at mid-summer scours away all the shadows and makes the landscape seem flat, but with each increment in the rotation of the earth, the light strikes from another angle and little hollows reveal themselves. The same landscape in winter, when the light comes more from the south, has yet

THE WEST SIDE.

a different texture, but it isn't necessary to wait for the seasons to change. The light shifts constantly throughout the day and the setting sun picks out high spots and pours shadows down into the low spots.

This doesn't even take into account the effects of weather. Drifting snow accumulates to give outlines to different terrain. A misty morning along the river can make the hilltops look like floating islands. Storm clouds can drag dark shadows across the slope or send bolts of lightning down to set fire to a haystack. Thirty-six hundred bales of hay burned one year when lightning struck a stack, blasting wire-tied bales in all directions.

That pretty much takes care of light and terrain, but there's more to see. There's that optimistic moment in the spring when you are fairly sure that you see a faint tinge of green in fields which have been brown with dried grass or (less often) white with our infrequent snow. Threads of darker green appear where low spots accumulate water faster. As anxiously as we await the greenup, there always seems to be a gap between when we say, "Is it beginning to look a little green?" and the moment we realize that it has already done so.

The gravelly, unirrigated hills will become a dusty green while the irrigated pasture turns a rich shade a manicured lawn fanatic might envy. (A lawn is merely a pasture that isn't being grazed.) Before that happens, however, a line of fire and smoke will often move across it, when dried grasses and weeds are burned in the irrigation ditches to allow water to flow freely. From time to time the stubble in the fields will burn too, highlighting the boundaries of the fields in deep, ashen black, which turns green again in a blink of the eye.

The fields and hills lack the geometric perfection of Midwest croplands, but the different shades of green define the land to the experienced eye.

Trees and shrubs along the river provide gray, brown and black during the winter. The twigs thicken in early spring as buds swell and at last, in little more than a fortnight, tiny golden leaves spread into pale green and shortly thereafter broad cottonwood leaves cover branches which conceal everything from orioles to eagles.

Summer goes by quickly, but we are forewarned of its end. A month before fall really arrives, one cottonwood tree will always allow the leaves on one of its branches to turn yellow or gold; yellow if the rainfall has been typical, gold or even orange if it has been a wet year. In the foreground, orange and red alder can be detected within the tangle of fading willow and birch.

Other colors make an occasional appearance: White, blue, green and red trucks, tractors the color of a pumpkin or classic yellow and green cross the fields. Orange canvas dams on the irrigation ditches stand out against the green. Sometimes a person or two passes by, barely visible on the distant road which runs along the big ditch. In summer, the black, yellow or golden dogs which often accompany them will be hidden by the tall grass, their presence only revealed by the actions of their humans throwing sticks for them to chase.

It is also hard to see the deer. Their coats take on the tired, gray-brown of late winter when the fields are nearly the same shade. They sport a richer, golden brown in late summer which is not so much the color of the fields as it is of the golden evening light. And the tall grasses of summer conceal them entirely.

Once you add livestock and wildlife to the scene, the variations are endless. Cattle, deer, elk, moose, foxes, horses, mules and coyotes can be seen from a distance. Up close, there are the rodents and insects. Birds fly by singly, in pairs and in flocks numbering into the hundreds.

A closer look at that little slice of land reveals burrows, indentations in the grass which tell of napping wildlife. A hollow may reveal a fawn —immobile despite your presence because its mother has told it to stay there.

And what about nighttime? Storms? Meteor showers? Deep, dark and bright moonlight?

A picture is said to be worth a thousand words. That patch of land is an ever-changing picture of incalculable worth.

How Do You Scratch Your Head?

Problem solvers have itchy heads.

As mentioned, I demonstrated blacksmithing in the 1935 shop on the ranch. Unfortunately, Johnny Grant's shop of the 1860s is long gone, but I gradually realized that a more modern shop might be an advantage; local men were still around who had worked in the existing shop. Short of a séance, there would have been no way of interviewing Grant's employees.

I had no hesitation in asking questions of former employees from the 1930s and 40s as well as visitors who labored in those years.

My questions were nearly always met with a frown of concentration and a furrowed brow. The arms were folded at waist level, then, gradually, one hand came up. Most often, the forefinger and thumb rubbed at the corners of the ponderer's mouth, the eyes narrowed, the head nodded and, "Ta Da!" The answer was there.

This was not an invariable routine. Sometimes the head bent forward, brows were raised, lips pursed and an index and middle finger gently scratched behind an ear.

Since most of our visitors are American or Canadian, it was some time before I realized that these were most often the mannerisms of people of British heritage.

German habits of thought were the next to emerge. We got a lot of visitors from Germany, and I noticed that they also started with the folded arms, but this was generally followed by an unfocused gaze and one hand stroking the throat.

Once I'd picked up on that difference, I began to keep track.

Even clean-shaven men of Arab or Jewish descent stroked their chins as if pulling at the beards common to their ancestors.

As an aside, arms folded high on the chest are a challenge and not a part of any useful thinking process.

Northern Europeans tended to indulge in a thoughtful three-fingered scratch just above the ear, although a Swedish visitor, questioned on his head-scratching technique demonstrated with a small, index finger circle just at the edge of his forehead. Before the suspicion could crystallize that he was telling me I was nuts, another Swedish visitor confirmed it as a habit of thought rather than insult. A French workman has a two-part head scratch. First, his hat is removed, then his other hand rubs or scratches the back of his head.

These national differences might be habits learned through observation but an intriguing possibility exists. I mentioned to an American couple that in the Netherlands, the thumb and middle finger are drawn down on either side of the nose. The husband jumped back, pointed an accusing finger at his wife and gloated, "Ha! That's what YOU do!"

Her ancestry, it transpired, was Dutch, but so distant that the habit might be a matter of genetics.

A second vote was cast for nature rather than nurture by a man who, despite having a Norwegian father and a German mother, rubbed the corners of his mouth, British style. He had been adopted as an infant, he revealed, and his birth parents were English and Welsh.

A thoughtful pulling down of three fingertips from lips to chin characterizes the Chinese and Japanese thought process. They seemed to be the only groups to skip the folded-arm stage of deep thought. They are also the only groups for whom I learned a specific reason for the gesture.

An exchange student from China mentioned ancient scrolls showing scholars and gods. They were often portrayed stroking their long beards. "We make that gesture in respect for them," my informant said. This, unfortunately, does not entirely resolve the question. Why did the scholars and gods stroke their chins?

Head scratching is most frequently a masculine habit. Regardless of national origin, a woman is most likely to place her thumb along her jawline with her index finger resting along her cheek and her middle, ring and little fingers covering her mouth. It is, perhaps, ungracious of me to suggest that she is thinking, "Well, I know what to do, but I'm certainly not going to try to tell him!"

There was a time in this country when a perplexing problem was said to cause a lot of head scratching.

Does that mean there were more thinkers in the past or that the general populace today is seldom called upon to do problem solving? Why do more men than women scratch their heads?

Does the head-scratching complete some sort of circuit between hand and brain? Which came first; thinking or scratching? Should our schools be teaching this specialized type of cogitation? And (perish the thought) do elementary school children have a symbiotic relationship with head lice, the former providing sustenance and the latter provoking thought?

Three Small Mysteries

FIRST MYSTERY: One morning, I pulled into the employee parking area, stopped, backed out and drove north a mile along the former Milwaukee Road right-of-way.

I didn't ask myself why. I just did it.

When I reached the boundary of the ranch, I found that cows belonging to our neighbor to the north had gotten out and were crossing back and forth across the still-active tracks of the adjacent Burlington Northern. They were enjoying the ample grass. They weren't in immediate danger, but were concealed in an area surrounded by cottonwood trees, and might well have remained there, unseen, until train time. I put them back in.

SECOND MYSTERY: On a hot July evening, after I moved to town, I realized I'd left my windbreaker in the blacksmith shop. I wouldn't need it, but nonetheless I got back in the truck and drove to the ranch to retrieve it.

When I stopped at the railroad crossing I saw smoke and flames from a fire along the tracks. We have become accustomed to such spot fires, triggered by sparks from a "flat wheel" on passing trains. It is very unusual for wind to blow from the northeast, but that was the case that evening. Though the smoke and flames were still low in an area of short, dry grasses, the wind was pushing them towards an area of cottonwood trees and tall, dead vegetation. From there it would have been little more than a few yards to historic ranch structures.

I had a shovel in the back of the truck. It only took a few minutes to cut enough of a fireline to slow it so I could get to a phone and call the fire department. They responded immediately and the fire was soon out.

THIRD MYSTERY: During calving season, the heifers (first-calf cows) were put up in a sheltered corral with access to stalls. Late one night I went up to check and one heifer had just started to calve. The emerging calf was positioned right and the heifer was doing her job. There was no problem that I could see. Yet, somehow, I felt there was a problem.

I was embarrassed to call my friend, Gene, for backup, but I did. Ten minutes later he arrived. The wind had come up. The sky was spitting ice chips and the tongue on the emerging calf was turning purple. Not good.

Gene pulled on the calf and I pulled on Gene and the calf emerged before the situation got critical, but I'll always wonder.

Why did I drive north that day?

Why did I go back for a jacket I didn't need?

Why did I call for help before I needed it?

A rancher friend told me, "The ranch called you."

I'd like to think so.

Pyroglyphics, Part 1

Livestock brands aren't just a fancy monogram you put on your cattle and horses. Your initials embroidered on a jacket or a towel may say, "This is mine." But to a cowboy, especially in the open range days, the brand showed legal ownership of a cow, and even indicated a philosophical ownership of the cowboy.

Perhaps that's overstating the case, but a cowboy rode for the brand. The brand showed what kind of hand you were—whether you rode for a good outfit or not. It told who your pals were. It described the range you rode and whether the chuckwagon cook was any good.

We were branding **Lazy G Hanging K** when I came to the ranch. Throughout the history of the ranch, Johnny Grant, Conrad Kohrs and Con Warren had used other brands. It didn't matter much to me whether we used an historic brand or not. I was not even aware of the legal requirements of branding until we put a brand on upside-down on one of our calves. This meant that instead of reading it **Lazy G Hanging K,** it would be read **Inverted (or Crazy) K Hanging Lazy G.**

It also meant we no longer owned that calf—at least not without jumping through some embarrassing legal hoops. The animal goes with the brand, and all the legal brands are registered with the state.

We had to call the brand inspector. He also happened to be the county sheriff. After he was nearly done laughing at us, he confirmed that the calf with the wrong brand was mothered up to a cow with our brand; an indication that it was probably our calf. Then he contacted the owners of the calf's brand. They could have taken possession, but they kindly did the paperwork acknowledging our ownership. We had to keep that paperwork to prove ownership until the calf was sold. Then we had to live down the error and nearly forty years later I can still hear the echoes of the sheriff hollering, "Say, you Grant-Kohrs folks brand any calves upside-down lately?"

COWBOYS BRAND DOGIES. COWCATS BRAND DOGGIES.

"No, Dave."

After a few years we acquired registration for one of the brands historically associated with the ranch, and we used it for a while, but it felt wrong. It wasn't "our" brand. It was the brand of a previous owner. I'd done a lot of research in the record vault of the Montana Department of Livestock by then, and was beginning to have a bit more understanding of what a brand really means to a rancher or a hired hand.

A cowboy didn't ride for the owner of the ranch. He didn't even ride for the foreman or wagon boss, though he acknowledged his authority. He rode for the brand.

When Conrad Kohrs bought Johnny Grant's ranch in 1866, he used Grant's [insert c hanging j] brand for a while, but soon started branding **CK**, as well as other brands relating to his various partnerships.

I suspect he felt about the **CK** as I felt about the **Lazy G Hanging K**. I didn't want a hand-me-down brand, no matter how historic. It's one thing to use a brand that's been in the family for decades. It's another to use a brand which has someone else's reputation attached to it.

Lazy G Hanging K that's us; Grant-Kohrs Ranch National Historic Site.

Going Back to High School

Sometimes I can't believe I voluntarily went back into a classroom. Admittedly, it was to teach a bit of blacksmithing to high school Ag classes. My school days were far behind me. I didn't realize high school still had a lot to teach me.

Frank Westfall was the Ag-Ed teacher and FFA advisor at Powell County High School for 27 years. Since 1986, Bill Lombardi has been the teacher/advisor. In my high school days, wood shop, mechanics and even drafting classes were only available to boys. Girls learned to cook and sew. Or as in my case didn't learn to cook and sew.

Teaching agriculture, like farming or ranching, is a year-round, round the clock commitment. This includes checking livestock at the school farm at any hour and in every weather, visiting students' homes to evaluate "SAE's" (supervised agricultural experiences) and accepting that summer vacation forms no part of an Ag teacher's life. The teachers have to be specially certified to teach the courses.

Frank had—and Bill has—the exceptional dedication necessary. I consider myself extremely fortunate to have been able to spend time with the Ag classes, even though I'm often jealous of all the things the students are learning. I frequently eavesdrop on Bill's lectures and demonstrations.

The first class I taught got off to a bit of a rough start. There were only boys, and I asked, "Do you ever have girls in Ag-Ed?"

"Yeah," replied a ranch kid who at sixteen already towered over me, "but we got rid of them."

I pointed out, fairly politely I thought, that someone who was not male was about to teach him. The humor was lost on him, however. A middle-aged female is not a "girl," certainly not in the eyes of a teenage boy.

Students could take Ag-Ed classes without joining FFA (Future Farmers of America), but FFA made the experience much better.

Title 9 passed in 1972, and suddenly schools receiving federal funding had to offer the same educational advantages to girls as to boys. It was not yet law when I was in school. I had never—outside Girl Scouts—been involved in anything involving teamwork, competition or sportsmanship. Admirable as he was, Frank resisted enrolling girls in FFA until 1977.

FFA students competed all over the state in public speaking, livestock judging, welding, silviculture and more. They worked together on projects as diverse as building a sign trailer for the highway department and constructing a greenhouse. They helped and counted on each other.

I hate judging people, but I learned a lot from FFA and 4H competitions.

Early on I was asked to help judge a speech competition. One of my shop students was in it. I knew him. He was so painfully shy that his face turned red and his throat closed up just trying to say hello. He stumbled and choked his way through his prepared speech while I tried to silently radiate help to him.

As I sat with the judges to rank the different speakers, he was scored very harshly and I protested, "But he's shy. He ought to get some credit just for standing up there!"

I was (ruthlessly, I thought) contradicted. He was to be judged by the criteria on his scorecard, not by any personal knowledge we might have.

I avoided any judging after that, but my fellow judges had it right. Four years later, that stammering student was the president of the Deer Lodge FFA, and his farewell speech before turning the gavel over to the incoming president was amazing: Confident, clear, and wise.

Having joined the National Park Service at a time when females weren't even allowed to wear the ranger uniform, I tended to be a bit militant about equality in shop class.

Two students usually worked at each forge, and I assigned boys and girls to different stations with impartiality. This, I finally learned, was a mistake.

The boys always seemed to take over, often doing the girls' projects for them.

Equality under the law notwithstanding, boys and girls learn differently. Boys like to just charge in; girls like to watch and assimilate before starting. Boys saw the hesitation as fear or inability and stepped in to help. The girls stepped out of their way.

The solution was simple. Team boys with boys and leave them to it. Team girls with girls and let them help each other.

IN 1932, THE DEER LODGE FFA WAS JUDGED THE TOP CHAPTER IN THE NATION, OUT OF AROUND 3,000 SCHOOLS WITH AGRICULTURE PROGRAMS.

That, at least, was the case in the early years, but I was able to watch changes through more than 20 years of classes. Law and time eroded some old gender stereotypes.

Still, I suspect it will take a few more generations before an eager male student fails to ask on the first day, "Are you going to teach us how to make a sword?"

Speaking of Swords:
AN ETHICAL DILEMMA

You can't work at a forge for years without having someone ask, "Can you make a sword?"

The answer has always been, is and always will be: "No."

Ever since an unfortunate incident in which I nicked my cousin with a steak knife when we were sword fighting, I have been aware that swords are not very friendly.

My punishment was poetic. My father had been teaching fencing at our community center (undoubtedly the inspiration for our foolish battle), and he put me in a class with kids five or six years older. Since I was seven or eight at the time, the size differential was considerable. I spent the entire time crouching on the floor.

By the time the Park Service put me in a Civil War uniform, complete with sword, I felt no inclination to draw it out to show visitors to the fort.

When I bought my house, there was a junk-filled log shop out back and among the Model A parts, scrap lumber and stray hardware, I found a bayonet.

The blade was long and slim, three-sided and designed for just one very unpleasant purpose. It was marked USΔ on the blade.

I checked with a military history buff who tentatively dated it to the 1870s. The Deer Lodge volunteers who fought at the 1877 Battle of the Big Hole were issued bayonets. This one would require a lot more research before it could even be speculated that it was of that era and had been issued to a volunteer, let alone that it had been at the battle.

"Provenance," however, was not the problem. The problem was that I hated the sight of it. As an historian, I recognized that it could have historical value. What did I owe history? Should I take it down to Big Hole Battlefield and see if it fit in with their Scope of Collections? (That's the policy which prevents us from accepting donations of totally inappropriate items, no matter how interesting the object or well-intentioned the donor.)

Two things finally decided me. First, I was repelled by the thought of putting it on display where—no matter how carefully labeled and interpreted—some war buff was bound to enthuse over its ability to disembowel people. Second, it came with the shop. I owned it. And since it was indisputably MINE, I could decide its fate.

It is no longer a bayonet.

Right or wrong from an historical point of view, I turned it into a candleholder. As I reshaped it around the horn of the anvil, I gave a fleeting thought to hammering swords into plowshares. I'll have to research that someday, because it has always seemed to me that a sword is too narrow and a plowshare too heavy and broad for that to be done successfully. I suppose, however, that depends on the sword, the plowshare—and the blacksmith.

The bayonet was so easy to reshape that it might have originally been intended to be a candleholder. The long blade curved around beautifully and the socket was precisely the right size to hold a slim taper. The only problem was the pointed tip. As I bent it over to form a handle, I felt a slight vibration which told me that bit of metal was close to fracturing.

I stopped right there and no one seems to notice that the handle needed to be shaped a little more.

My final bit of historical vandalism waited a couple of years as I tried to find an engraver who could work on the concave "blood channels" of the blade. Modern engravers, however, don't hand-engrave and their computer-driven programs couldn't cope, so I gave up and did it myself.

It's finished now. The engraving around the base says, "It is better to shed light."

The Past Can Be Puzzling

Visitors often say, "I was born one hundred years too late."

Quite apart from the fact that I would now be long dead, I have no desire to have been a rugged pioneer.

And the important thing, it seems to me, is that all those simpler things people pine for still exist.

Out behind the ranch house are several crudely-made wooden roping calves. Visitors young and old are welcome to pick up a lasso and try roping. Sometimes an employee is available to help; sometimes they're on their own. I preferred to just give them a couple of pointers and walk away. They didn't need the pressure of trying to measure up to my rudimentary teaching.

It's a fact: You can't talk on a cell phone and throw a rope at the same time. Phones and other electronic gadgets are put away while toddlers to teenagers get totally wrapped up in one of the lowest-tech activities some of them have ever seen.

Long before I knew what a blacksmith was, blacksmith puzzles held a special fascination for me. My grandfather had given me the simplest one—two looped nails slipped together—when I was about five, and he said, "If you have to force it, you're doing it wrong." I eventually solved the puzzle and, being a small child, lost it, but the lesson stayed with me for life. This, perhaps, was due to the fact that shortly after I'd worked the puzzle, I was trying to screw the lid back on a peanut butter jar. It was crooked and I was about to try force when my grandfather's words came back to me. After all, it had come off easily enough. It should go back on the same way. I unscrewed it, placed it properly and put it on easily. Admittedly, the rule doesn't apply in all cases. For example, I have to force myself to do housework, but that doesn't necessarily make it wrong.

In the blacksmith shop I made iron puzzles and wooden puzzles visitors could play with. I loved to pass out four or five different puzzles to a family. At the start, they'd all work in separate parts of the shop, but after a few minutes they unconsciously began to draw together. They didn't seem to be doing so to help each other. Each one was absorbed in solving the puzzle in hand. They seemed, however, to appreciate that the rest of the "team" was with them.

After a while, as a puzzle was solved, I would have to step in briefly to stop them from showing each other the answers too soon.

ARE KIDS BORN KNOWING HOW TO PLAY CHECKERS?

One puzzle looked deceptively simple. It seemed obvious how it should work. Surprisingly, there were half a dozen wrong ways to try. Perhaps it was unsporting of me, but when an impatient parent tried to hurry a kid along with the traditional two-nail puzzle, I provided the tricky one to the adult. The kid usually solved the nail puzzle first.

It was a pleasant surprise when I found most middle school students weren't totally addicted to finding answers on the internet. When they'd ask me for the solution to a puzzle, I might give another clue, but would occasionally add, "Of course, you could probably find the solution on a computer, but what would you have learned?" They instantly replied, "Nothing!"

"That's right!" I'd congratulate them. "You already know how to find answers on the computer. This is about finding the answer in your own head."

There was also a hand pump set in a water-filled barrel in the shop. Introducing a wide-eyed three-year-old to the magic of pumping water was an amazing sight. They were often too young to understand why they were being asked to raise and lower the handle. After two or three tries, water started to come out, but they still look perplexed.

PUMPING WATER WAS A GRAND ADVENTURE FOR YOUNG KIDS.

Then came the moment when they made the connection: Their eyes widened. They looked up with an incredulous smile: They were making the water come out!

After that their parents couldn't drag them away. Sometimes the little ones would even slip away from their parents to race back to the shop and pump some more water.

Kids roping the wooden steers pleaded to try "just one more time!"

I didn't kid myself. They weren't going to toss out their electronic games just because they'd tossed a lasso at a wooden calf, separated a couple of twisted nails or worked a pump handle, but it was a start.

Happily, iron and wood puzzles are cheap and easy to make. When I saw or sensed real interest, I gave the visitors a puzzle to take along. It won't wear out. It won't break down. And the only thing which will need to stay charged is the visitor's curiosity.

Pyroglyphics, Part II

I never branded one of our animals, though I made a lot of our branding irons in the blacksmith shop. In fact, apart from the calf that escaped my hold back in 1976, the only cattle I've branded belonged to Montana State Prison. The prison encompasses 38,000 acres and includes both a dairy and beef cow operation. One year (I'll spare you the unhappy details), the inmates who normally did the ranch work were not available. I agreed to help, and though I did brand a few cows, my main job was to sort and hand out the irons which were being heated by propane.

Each cow was receiving an individual four-number brand—a herd number rather than the MP brand registered with the State. The MP brand had already been applied when the animal was a calf. The cows we were branding that day were becoming part of the breeding herd, and as such needed individual identification.

There were several irons in the fire: 0, 1, 2, 3, 4, 5, 6, 7, 8. If you're wondering where the 9 was, it was simply the 6 turned upside-down. Or perhaps, the 6 was really the 9. Whichever, it was important to hand the irons out in the right order or the cow would get the wrong number.

It was a hot day, and I hadn't been smart enough to keep my hat on. Standing by a hot fire with the sun beating down, I realized I was getting a little woozy. I turned away from the fire and took a few steps before fainting flat on my face in a corral which, to be frank, was a little untidy due to the passage of a number of stressed bovines.

When I came to, I poured some cold water on the back of my head, put my hat back on and got back to work. I couldn't have been away long: We were only a couple of numbers down the line.

We finally broke for a meal and I learned another fact of ranching. My absence had scarcely been noted, and there was general agreement that the rest of the crew thought I was just taking a break.

That's ranching for you. Right or wrong, wrecks are just part of the job, although I suspect a prison ranch may be a bit more "macho" than most.

Though my help wasn't really needed to brand the cows, I was allowed to do so. I wanted to learn the "mechanics," and the crew let me.

BRANDING AT RANCH TWO, MONTANA STATE PRISON.

It had seemed to me that the iron stayed on the animal for an inordinate length of time, but the color of the smoke was the key in knowing when to pull it away.

The first, pungent smoke was yellow. That was the hair burning. There is a lot of moisture in a live animal (I believe humans are purported to be about 75% water) and as soon as the iron got through the hair and touched the skin, the yellow smoke turned to white steam. A cow's side isn't flat as a board, so the iron pressed unevenly. This meant one area might still be smoking when the rest was steaming. No problem: The iron was simply rocked towards the smoky side until it was steaming too and then it was pulled away.

These days, Montana law requires the brand figures to be a quarter of an inch wide at the base. Some of the historic irons at Grant-Kohrs are practically knife sharp. They were not intended to cut, however. With a great deal more experience than most folks have these days, those old-time cowboys knew just how to touch the iron to the animal. That made a very clear brand, killing hair follicles so the surrounding hair would grow back in a distinct pattern.

Keeping the heat at the right temperature had probably been more challenging when it came from a wood fire in bright sunlight on the open range. Too, on a big roundup, the individual irons of several different ranches might be heating in the same fire, because different herds mingled on common ranges. When a calf was roped, its mother would naturally come to protect it. The brand on the cow would be read and called out to the man tending the heating irons and the appropriate iron would be handed up and applied to the calf.

Visitors are often surprised that cattle are still branded in Montana, but this is a big state and the cattle roam over thousands of acres. It's not unusual for a break in a fence to allow herds to mix, just as in the old days. Cattle rustling is still a problem, and I have learned several ways to steal cattle.

Have I stolen any?

This is a reminiscence, not a confession.

Con and Beechnut: Old Friends

That is a misleading title. I don't know if Con and Beechnut ever met. I don't know if Con liked cats or Beechnut liked ranchers. They were both friends of mine.

Beechnut, my number one sidekick, lived with me, five younger cats, cattle, horses, poultry and miscellaneous wildlife. No other Park employees lived on the ranch at that time. Each evening I'd go for a walk, trailed by, accompanied by or sometimes led by the cats and an occassional skunk. Cats are wiser than dogs when it comes to skunks. They coexist fairly well, and on rare occasions we were even joined by a confused or lonely skunk.

One calm summer evening our path led us by a horse trough and the cats hopped up on the adjacent fence to take a drink. With their hind feet firmly on a fence rail, they stretched their front paws across to the rim of the trough. The younger cats had no trouble, but Beechnut, at the extraordinary age (for a ranch cat) of twenty-one, was getting rather stiff. Thinking to help him, I picked him up and sat on the fence with him in my lap.

Though he was usually a real guzzler, he pushed away, jumped down and stalked off.

"Hmph!" I thought. "That's all the thanks I get."

We continued on our walk and after a bit I noticed that Beechnut wasn't with us. Retracing our route, I found him perched on the fence, drinking from the trough. He had been thirsty, all right, but he was quite capable of getting his own drink, thank you.

91

The next afternoon, I stopped by Con's house for coffee. He was eighty-three then, and in rather fragile health. Over recent months our coffee ritual had changed bit by bit. At first, he would get out the cups, set the water on to boil, fix my coffee (black, with two ice cubes) and his (with plenty of cream and sugar).

But it seemed so difficult for him! Gradually, I started to "help." I'd set the kettle on while he got the cups. Next, I took over the making of my coffee. Why should he have to wrestle with the ice tray when I was the one who used the ice? It was so easy to take over the whole thing: "You just set there and relax. I can do it."

He was a lot more polite than Beechnut, but no less independent at heart, and what did I do to that independence through my well-meaning efforts?

OK, it was hard for him to lift that heavy kettle, but did that mean he shouldn't have lifted it? I took the role of host away from him in his own home! I as much as told him, "You can't handle something as simple as fixing a couple of cups of coffee," even though that wasn't true at the time. And what was my hurry? Was I there for a free cup of coffee or to visit a friend?

As I sat there that day, I felt my face flame with embarrassment. Was it too late? Could I return his independence to him?

I was able to share some of the coffee-making ritual with him after that, though I continued to do some of it, because I couldn't figure out how to completely reverse the trend I'd started.

Over the next two years he became frailer and finally even the walk to the kitchen table became too much, but I never again made the mistake of leaping up to do something for him that he could do for himself.

And I didn't help Beechnut either, until he asked me to. As any cat lover knows, a cat is perfectly capable of asking for help.

So is a human.

Animal Training

This is not about training animals. It is about being trained by animals. They are patient and persevering and have taught me a lot.

Pierre taught me one of my best tricks. She was a long-haired orange kitten who announced her arrival at the ranch with the loudest "meow" I've ever heard. It sounded like a hysterical parrot shrieking "Pierre! Pierre!"

She was not alone. Her brother, a long-haired gray cat strayed in with her, but he was silent and stoic. He was named Trout.

As soon as they were captured and old enough, they were "repaired" as we delicately phrase it. They joined the rest of the cat herd without squabbling and went with us on our evening walks.

I began to notice that Pierre had an odd habit. As she'd approach me, she would drop and roll when she was about 10 feet away. Just for fun, as she approached the "drop zone" I'd say "Hey Pierre! Roll over!" and, of course, she would. People were astonished. "How did you teach her that?" they would exclaim. "Oh," I would nonchalantly reply, "you just have to be patient."

Pierre had been patient. It took her a while to teach me to say, "Roll over," at just the right moment, but I finally got it.

It was unsporting of me, but I used the same ploy with any ranch cats who followed us to the front door of the ranch house when I was giving a tour. Cats are not allowed in the house, although I did once give a tour with my huge parka pockets full of ranch kittens.

The cats knew the routine, so as soon as we entered the house, the cats would circle around the back porch and greet us as we came out. So, before we went inside, I'd say, "Meet us on the back porch," and they would. "What smart cats!" the visitors would exclaim. "They understood you!"

Inkle, a prime mouser for over 20 years was outside the blacksmith shop one day. I put a bit of cat food on a board for her and a visitor laughed and said, "Isn't she supposed to be mousing?" Inkle immediately trotted off and within a minute or so returned with a fat vole which she dropped beside the board before going back to the dry food." Though I was as surprised as the visitor, I casually explained that she preferred to trade the rodents for things that didn't get stuck in her teeth." It had to be a coincidence. I'm not even certain what the lesson was unless it was to be careful what you say around a champion mouser.

It hasn't only been ranch critters who have educated me. Halfway around the world, I received an important lesson from a wasp. I was sitting at an outdoor café in Strasbourg, France, drinking apricot juice. A wasp kept hovering around my glass. As often as I waved it away, it came back. I kept the lid on the bottle of juice so it wouldn't crawl into it, but each time I took the lid off, it returned. Finally, common sense prevailed. I poured a bit of juice into the lid and set it on the table. The wasp settled down to enjoy its few drops of nectar and I drank the rest.

The lesson may not have been intentional, but the message was clear: Share.

Banging on an empty water trough doesn't make water come on, but horses know that banging on it will bring a human over to fill it. Well, actually, it will bring a human to fix or replace it. Feeding the livestock and checking their water are daily chores, and an empty trough usually means that it has either sprung a leak or—if it's one that tops itself off—the float valve or some other mechanism isn't working. Since that doesn't happen often, I'd like to know how they figured out how to notify us. We certainly never said, "Bang on the trough if you need water." They, however, clearly taught us to respond.

Thomas, an excellent cow pony, taught me to have the halter unbuckled and ready before I entered his corral. As soon as he saw it he would thrust his head into it. If it wasn't ready there was an awkward tangle of straps and lead rope and I could nearly feel his contempt. It took a couple of tangles before he had me trained.

Horses appreciate a little courtesy. Charging out to catch them, rope in hand, is nearly guaranteed to make even the most cooperative horse turn away (with the exception of Thomas, of course). An experienced ranch hand could have explained this to me, but the horses did the job for themselves. A direct approach nearly always caused them to turn away. A patient, quiet and oblique approach nearly always resulted in a successful catch.

Lucky, a saddle horse, was an exception on the opposite end of the spectrum from Thomas. When he was in open pasture, the only way to catch him was to take another horse, saddle up, ride out and rope him. Of course, by then you were already on a horse, so there wasn't much point in catching him. It didn't take long before we stopped trying. I can almost imagine him thinking, "Quick learners."

Café College

Morning coffee at the local café is like a seminar. My coffee cronies have expertise in saddlemaking, wheelwrighting, logging, government packing, horseshoeing, gunsmithing, cattle ranching, irrigating, construction, mechanics, electrical wiring and ironwork. A week rarely goes by without someone raising a question about any of these topics. Since I know the least, I'm often the questioner.

Sometimes the discussion is general. Sometimes an expert gives a simple and direct answer. Sometimes a pen comes out. Those are the best times.

Knife, fork and spoon are unceremoniously dumped onto the table as the napkin they had rested on is turned into a very special and specific textbook.

The lesson starts off with a question. "Say," the conversation begins, "I found a really strange piece of iron out in the old barn the other day. Maybe one of you might know what it is."

"What did it look like?" asks one of the group.

"Well, it was about six inches long and it had a handle at the top and a sort of hook at the bottom and a thing that swiveled out. I think it was handmade."

The fact that there isn't enough information yet to make an informed guess doesn't stop us. The questions simply continue:

Q. *Was it some sort of hay hook?*
A. I don't think so. The part that swiveled would get caught in the bale.

Q. *Did it have any writing on it?*
A. No

Q. *What made it look hand forged?*
A. Well, it was really well done, but the handle at the top was sort of folded over a couple of times like it had been sort of forge welded.

It has now become apparent that we probably don't have a good mental picture, so we push the napkin and a pen towards the speaker and say, "Can you draw it?"

Sometimes a sketch provides the necessary clue. Sometimes a few scratched-out attempts lead to "I can't draw!"

I would have made a lousy police sketch artist, but after a couple of decades at Café College, I'm getting better at drawing people's mysterious bits of iron. If I ask enough questions, I can usually come up with something vaguely resembling the object. Of course, if it has any moving parts, the matter is considerably complicated.

"Yeah, it was sort of like that" is progress but may not lead to a solution.

There are outside experts to consult—outside our table, that is. One of our group notices folks sitting at another table who work at the mill, a ranch, the county road crew, etc. "Say, you guys," ("Say" being the proper word to forewarn people they are about to be questioned) "do any of you guys know what this is?"

The sketch is passed around and discussed by anyone likely to have an informed opinion. Maybe the answer is forthcoming, or maybe the quest continues.

"Why don't you bring it in tomorrow?"

"I would if I could remember where I put it!"

This is often the death knell, but if the mystery object can be brought in there are several options.

- Everyone says, "I don't have any idea."

- Someone says, "It looks like one of those old Doohickeys, except for the handle and the swivel part."

A TYPICAL CAFÉ COLLEGE NAPKIN.

- Someone says, "Oh! I know what that is! It's an old Whimwam. We used them when I was a kid."

A bibliography and appendix for this unique textbook is compiled when someone says, "Say, you know who might recognize this is old Whatshisname, you know, Marie's sister's husband."

This leads to the extremely complex genealogical information common to small town and to digressions into unrelated stories which I, as the usual napkin archivist, add to the sketch.

By the time the answer is found or the investigation abandoned, the napkin will have names, sketches, and anecdotes ranging from the practice of roasting potatoes and apples wrapped in foil in hot clinkers pitched out of old steam engines to the proper technique for turning down the pilot light in a gas stove.

The drawback in this type of specialized text is that it is hard to keep it accessible for future researchers.

The Dewey decimal system is of very little use in putting a paper napkin textbook on the right shelf.

Hun and Johnny

Thanks to the unexpected encouragement of a couple of friends, I found myself spending four great years on the Montana Committee for the Humanities (now called Humanities Montana). I hadn't thought myself qualified, but it turned out that the committee considered all humans to be part of humanity (imagine that!) and members came from all over the state and all kinds of backgrounds. Four times a year we would be sent a stack of grant requests to read. We'd meet, and the requests would be debated, sometimes quite vigorously, for the first day, and voted on the second day.

Grants funded an incredible range of subjects, among them the speakers who gave public programs at libraries across Montana, recording equipment used to capture native languages before they were lost and assistance in producing a film about women bronc riders in the early half of the 1900s.

Near the end of my term I attended a couple of national conventions. The first was in Washington D.C. and although we had been given a group rate which was not unreasonable, the nation's capitol is still a bit rich for my blood.

As was my habit, I had chatted with many of the hotel staff. They came from all over the world, refugees from whatever chaos had enveloped their homelands.

One evening at about 7:30 there was a tap at the door. A tiny maid stood there, ready to turn the bed down and put a piece of chocolate (wrapped in a fancy box) on my pillow. I said, "Thank you, I think I can turn the bed down."

She handed me two pieces of the chocolate and I asked where she was from.

This is where the narrative gets sticky, because it was very complex.

Her name was Hun. As we talked, I learned that in 1974 she had lived in Cambodia and "they" (the Khmer Rouge) were killing her family, along with an estimated three million other people. One sister had already been killed. She and her father fled to Thailand with her three children. The children died. Her husband and another sister fled to Laos. She already had a sister in America and after some time (perhaps five years) she came to America to join her. Her husband survived and came too. They "got" more children. (I wasn't sure if she meant they had more children or adopted them.)

I gestured at my room which had seemed embarrassingly large and asked, "What do you think of all…this?" Her reply made me feel about two feet tall.

Hun answered that it was good because it gave her a job that supported her family. Then she gave a start and said she had to do the other rooms. The door closed behind her but I had scarcely turned away when there was another knock. I opened the door and she thrust several chocolates at me and hurried down the hall. I stood there with my hands full of frivolous chocolate boxes and my eyes full of tears.

The second incident happened shortly after 9/11 at a meeting of the same group in a mid-western state.

Of course, the recent horrors were very much on everyone's mind. This time we stayed at a nice but not embarrassing hotel. Coming down in the elevator, I noticed the nametag of the hotel employee standing opposite me. The first name was Johnny. The last was an Arab name. He saw that I had read the name and he quickly ducked his head down.

On the ground floor, the door opened and he put out an arm to keep it open for me. As I got out, I said "shukren" which is Arabic for "thank you" and which was almost the only word I knew.

He murmured a one-word reply, but what stunned me was that his eyes had filled with tears, nothing so dramatic as weeping, but a hint of the nature of his experiences since 9/11.

It occurred to me that these two incidents had much in common and that it could not be mere coincidence. I could not have met with the only two hotel employees with histories which could make me cry.

Hun's experiences occurred in Asia and she and her family fled to America. I don't know what Johnny's experiences may have been, but it would be beyond tragic if the aftermath of 9/11 turned our own country into a place for him to flee.

It was ironic. After four years on the committee, I realized I had a lot to learn about the humanities—and humanity.

The House Always Wins

It isn't difficult to make comparisons between gambling and agriculture.

If you're old enough, you'll remember when "one-armed bandits" were slot machines into which you dropped actual coins. Then you pulled a lever down and reels inside the machines spun around, eventually clicking to a stop, one at a time. There was the excitement of pulling the handle, eagerly watching the spin and click of the reels, and, if you won, the clatter of your winnings falling into a metal bowl. Now, of course, the machines are electronic. They have replaced that old excitement with speed and synthesized sounds. Since the house always wins eventually, all that speed just enables you to lose your money faster.

I have no illusions about my luck. I was given a toy poker game 20 years ago and have kept track of most of my losses ever since. I quickly realized that there was no point in keeping track of my winnings. There weren't any. In a score of years, I have lost an imaginary $70,000 and three AA batteries.

Imagine that, instead of dropping your money into a slot machine, you dropped it into a furrow, covered it with soil, watered and weeded it and watched as your "winnings" emerged. If nature cooperated with a judicious amount of sun and rain, if no infestation of insects devoured the plants, if an untimely frost or hailstorm didn't flatten the fields, you might hit the jackpot. It takes guts to bury your future.

Imagine that, instead of investing in the stock market, you invest in livestock. You bet that your cows will have healthy calves, that scours (you don't want the details) and predators don't take a cut, that the livestock market will be good, that fuel prices to transport the animals to feedlots won't be too high, that prices of feed to "finish" the cattle will be low, that public tastes will support sales of beef.

It may be that in agriculture, as in gambling, the house always wins, but there's a difference. The "house" in agriculture is largely owned by nature, and nature generously takes in partners. Establishing a partnership with nature means improving the odds in your favor.

When I moved to the ranch, I'd go eagerly through the seed catalogs and order all manner of starts for plants listed as "suitable" for the Deer Lodge Valley. Working with nature is not the same as working with a seed catalog. A pretty photo of some fruit or vegetable growing enthusiastically in a greenhouse has nothing to do with gardening success.

I was betting against the house. Generally, my optimistically-ordered plants froze. That seemed unfair, as there were folks in town who had success with the same plants. It was some time before I recognized that the numerous streams flowing through low areas at the ranch made the land 10 degrees or more colder than higher, drier areas and I revised our temperature range from the catalog's cheerful "Zones 3 and 4" to "Zone 0 to 2."

There is nothing wrong with optimism. Pessimism, on the other hand, is not useful. Optimism needs to be balanced with realism and knowledge. Pessimists would never plant anything.

Eventually I learned which plants were really suited to the area, but my trials were not over. Dirt is not simply dirt. Different plants require different soils. Also, it is not wise to put plants which grow tall on the sunny side of plants which need lots of sun. Putting plants which require lots of water next to plants which need to dry out between watering is also silly.

Sometimes the plants themselves put me straight. I kept planting raspberries in the wrong places. They didn't die, but they produced no berries. Then, one spring, I found they had sent runners around the corner of the shop and colonized a spot better suited to their needs. Had I dug them out when they failed to produce, I wouldn't be able to enjoy my annual abundant raspberry crop today.

At the local extension office, I learned that I should give new plants three years before either giving up on them or moving them. Trees and shrubs that barely struggle along for a couple of years suddenly take off the third year. I heard this was because they spent the previous years putting down roots, but I think they just bide their time while they make sure I'm going to remember to water them and won't carelessly mow them down.

I learned that working with nature means knowing the land, the soils, the availability of water, the most appropriate plants or livestock, weather patterns and much more. Once I realized that simply *wanting* to grow something didn't mean it was going to grow, I began to appreciate the accumulated wisdom of the people who had lived here for years.

This is "meat and potatoes country." That's what grows here.

For many ranch families the result of a successful partnership with nature isn't necessarily a pile of cash. It's being able to stay and work land which has been in the family for generations, and it's knowing the next generation will be able to do the same. The payoff for the rest of society is food. And the ranchers have one advantage: They're playing for the house.

Still, I tend to sympathize with a story that's been around for a long time. A rancher is asked, "What would you do if you had a couple of million dollars?" The rancher ponders for a bit, then replies, "Well…I'd probably keep on ranching…until it was all gone."

My Favorite Breed

Notwithstanding my pride in my native Montanan status, I knew very little about cattle when I transferred to the ranch. I still hold a bit of a grudge against a cattle buyer I questioned back the '80s. It was a simple question about the morning stock report, and he just shook his head and said, "You just sort of have to be born to it."

Nonsense!

On the other hand, there have been a lot of changes in the last four decades, and even if I'd grown up in the business there would still have been a lot to learn.

When I arrived in the mid-70s, Herefords were the breed to beat, as they had been for nearly three quarters of a century. There were beginning to be a lot of "black baldies;" a Hereford/Angus crossbreed which produced a black calf with a white face. Then, insidiously, pastures were invaded by all sorts of breeds, many from Europe. More than 240 breeds of cattle are recognized world-wide.

A rancher's preference for a particular breed is similar to a sports fan's preference for a particular team.

I recall standing in front of the ranch house one day with a group of visitors when a bellow announced the arrival of a white bull with black ears, nose and eyes. We dove for the front door as he trotted up the path. He turned out to be an Italian breed, a White Park. When his owner arrived to reclaim him, I asked, "Why do you raise White Parks?" "Cause they're pretty," was his laconic reply.

There was one justification for the increasing preference for Angus. Herefords of that era were light-pigmented, and subject to sunburned bags (udders) and cancer eye. Time and careful breeding have now gone a long way towards solving that particular problem. Meanwhile, dark-pigmented Angus cattle took over the range. Then clever marketing brought Angus cattle to the public as somehow superior in all ways. Ironically, branded cattle were being "branded" again in the world of advertising.

Personally, I love the Herefords. One of my favorite sights at the tag end of winter is a line of crisp, green hay on the snowy ground up near Gold Creek, with white-faced calves curled up on the hay like baby birds in a nest. Their faces are whiter than the snow.

Gold Creek is famed for being the site of an early confirmed discovery of gold in what was then part of Idaho Territory. It's 17 miles from the ranch, I had the good fortune to help with winter feeding and calving there during the winter of 1988. The temperature ranged between 30 and 40 below for three weeks at the height of calving.

As a result, I got pretty good at playing cribbage. We had to check the herd constantly. In order to stay awake between checks, we played cribbage, ate gingersnaps and drank gallons of coffee. A wet, newborn calf hitting frozen ground could freeze to it in minutes. The ranch house kitchen became a temporary home to a number of calves. Unless they were weak, they only had to stay in until they were dry and had a belly full of the first milk called colostrum, which is rich in antibodies.

It was a challenging three weeks. At one point, my light truck was the only vehicle which would run, despite the fact that it was two-wheel drive and had no block heater. It's quite an adventure feeding 450 hungry cows out of a small pickup. Dave used an ice axe to get to the top of the haystack and tip a dozen bales down onto the truck at a time. We didn't have to unload the bales. Each time I drove into the pasture a shark-like feeding frenzy broke out as the cows ripped the bales off the truck and tore them open. The bales were tied with wire, and we'd walk back over the pasture and gather the wires as soon as it was safe to do so.

Despite the cold (and partially due to the truck's heater) it wasn't too bad, but when I got home one afternoon, I counted the clothes I removed, and (not counting my belt) I had been wearing 21 items, including double layers of socks, long johns, jeans, sweat pants, two shirts, a wool vest, jacket, silk scarf, wool scarf and a lot more.

Archetypal cowboy Teddy "Blue" Abbott rode for Conrad Kohrs during the legendary hard winter of 1886-1887. In his autobiography, *We Pointed Them North,* he described how he survived the weather. "I wore two pairs of wool socks, a pair of moccasins, a pair of Dutch socks that came up to the knees, a pair of government overshoes, two suits of heavy underwear, pants, overalls, chaps, and a big heavy shirt. I got a pair of woman's stockings and cut the feet out and made sleeves. I wore wool gloves, and great big heavy mittens, a blanket-lined sourdough overcoat, and a great big sealskin cap. That way I kept warm enough, but not any too warm."

CHAPS AND A NECKERCHIEF AREN'T REGULATION, BUT COMMON
SENSE IS WHEN A SUDDEN SPRING SNOWSTORM DESCENDS.
A FRIENDLY DRAFT HORSE MAKES A GOOD WINDBREAK FOR GLEN.

In severe weather we had to do some doctoring of sick calves, and my preference for Herefords increased. It was a "commercial" herd, meaning that several different breeds were represented. We didn't need to doctor many Hereford calves, but when we did, the mothers stood by, watching to see that the calves weren't being hurt. They knew us. We were the people who fed them.

Angus cows, on the other hand, just wanted to kill us. Conventional wisdom has it that a cow won't run over her own calf to get to you, so if you keep the calf between you, you can give it a shot or a pill in comparative safety. Just for the record, an Angus cow will run over her own calf.

Now, there are ranchers who will say, "That's a good mother. She'll always bring home a calf." They have a point, but I'd rather get home alive myself.

There are reasons for breed preferences. Some breeds are more tolerant of heat. Some do better in rough country. Still, I know that my love of Herefords has as much to do with their history and my own Hereford bottle calves as their excellence. It's called being "barn blind."

A preference with even less justification has to do with the gender of a beef animal.

Pete had been a butcher for many years and told me with obvious contempt about a man who would place an order and say, "And don't give me any heifer beef, I know the difference!" Finding the man's attitude insulting, Pete took considerable satisfaction in giving the man nothing but heifer beef.

Ms. Know-It-All—NOT!

OK. I know a lot about the history of the ranch. After about 40 years I'd better! The problem is that I also know that I *don't* know a lot.

Grant-Kohrs Ranch National Historic Site was created to tell the story of the open range cattle era: all of it. That includes the story of the investors from back East and England and France. It includes the ranches that failed. It is the story of all the western ranches, extending down into Texas and out as far as Hawaii. It reaches back in time to the Spanish, to Columbus and to the Portuguese who brought cattle to the New World in the early 1500s. It includes the influence of open range cattlemen on the reduction of Indian reservations and the tribes which turned to stock-raising as their traditional ways of life were torn away.

And how about the stockyards in Chicago? The monopolies of the "Big Four" meat packers?

Just off the top of my head, a "know-it-all" at Grant-Kohrs Ranch would need to know about cowboys, railroads, horses, cattle, water—including water law and irrigating, grasses, tribal history, wagons, chuckwagon cooks, fences, veterinary care, trail drives, roundups, homesteads, Victorian furniture, eastern stockyards, barn cats, noxious weeds, wildflowers, tools, birds, seasons, blacksmithing, log construction, wildlife, farming, poultry, horse tack, from saddles to bridles to harness, oxen, mules, spurs, lariats, crops, immigration, horseshoeing, turn plate roofing, post-on-sill construction, politics, contemporary stock raising, farm machinery, brands and brand law, weather, including the effect of lightning on nitrogen availability—and that's just a start.

I'm not embarrassed to admit that I don't know it all. When a visitor came up with a question and I didn't have the answer, it was actually quite easy to say, "Beats the heck out of me! But I'll try to find out." That is, in fact, one of the best things about the ranch: There is no end to the things I have yet to learn.

What's embarrassing is being given credit for knowing more than I know. And I'm embarrassed to find out that something I "know" isn't true. Sometimes my source was wrong. Sometimes my source was lying. Sometimes I misunderstood or misremembered or simply wanted it to be true—and sometimes the truth changes.

One example should suffice: For years, we told visitors, particularly youngsters, the story of the cabinet bed in Johney Bielenberg's bedroom. We had, after all, learned it from a descendant, so it had to be true. As we were told the story, Johney was a bachelor and more comfortable with the hired hands than the society usually entertained in the ranch house. When he'd leave his room messy, the mess could be tossed on the bed and folded out of sight. Funny story. Not, apparently, true.

It was years before we really took note of a couple of oral histories from other relatives who stated very clearly that "Uncle Johney was a fastidious old bachelor who kept his room as neat as a pin." Oops.

When we tell a story like that, it can come back to haunt us, years after we've corrected ourselves. It happens like this: A new employee gives a house tour to visitors who have heard the old story. They share it with the new ranger who, finding it amusing, tells it to subsequent groups and a whole new generation of visitors carries it away—and brings it back.

The cabinet bed: One piece of furniture in one room of a 23-room house on a ranch with two houses, a bunkhouse, and dozens of other buildings in one part of a valley in one Western state…

Nobody knows it all.

Cast of Characters

Rangers who give walks and talks are called "interpreters," because we can, in theory, interpret complex and arcane subjects to the visitors. It's too tempting to make this our own personal interpretation. Many years of research about the key figures in the history of Grant-Kohrs Ranch tempt me to think I almost know them. I don't, of course.

I have a strong prejudice against role-playing because I'm not even certain I could portray myself honestly, and what makes me think I could portray someone who was raised in an entirely different culture and died over a century ago?

Take Quarra Grant, for example. A member of the Bannock tribe and related to Sacajawea of the Lewis and Clark Expedition, Quarra is remembered with some detail in Grant's memoir. Born around 1840, she bore the first of their six children in 1856. She spoke French, English and several Indian languages. When the 1862 portion of the Grant-Kohrs ranch house was built, Grant noted that she "showed a wonderful skill in taking up all the ways of white women. She could make my clothing as well as a tailor, and in every way, she was a capable wife, as bright and as merry as a lark; and she was very proud of our new house." He also noted that she could ride horses many men could not.

What a happy picture! It leaves out the presence, perhaps intermittent, of his other wives, and to maintain the picture the tale must stop short of the fear Quarra expressed of being left behind when Grant decided to return to Canada (though he assured her he would not do so). To his credit, Grant did her full justice, and his recounting of her death from tuberculosis while he was finding them a new home in Manitoba can reduce many visitors to tears.

I don't want to reduce visitors to tears and will have to confess that I have told the happy half of her story many more times than I have carried the tale to its truly bitter end. It's truth, as far as it goes, but it isn't honest.

Kohrs' half-brother, Johney Bielenberg was his partner in their vast ranching operation. It is convenient to portray Kohrs as the power in the partnership and Bielenberg as little more than a foreman, overseeing the day to day operation of the ranch. But of hundreds of copies of letters in our archives about the cattle operation of Kohrs & Bielenberg during their most prosperous years, nearly all were written by Johney. It was Johney who, in 1901, wrote prophetically that "Herefords are the coming breed for Montana." He was a territorial and state senator and was on the board of the Montana Stockgrowers. In our efforts to tell the story simply, we could easily do a disservice to complex characters.

Sometimes little more than a sentence or two can create a believable character. Con Warren told me that his great-uncle, Johney Bielenberg, told him that "any man who would lie, cheat or steal isn't a man." As a child, Con felt he could never measure up. Was Con's recollection telling me more about Johney, or young Con?

The history of Grant-Kohrs Ranch is so fantastic and the main "characters" so exceptional and the records so complete that it would be possible to reconstruct much of the ranch story year by year. But "much of the story" isn't the whole story.

Along bunkhouse row is a stall which was converted into a room for the chore boy. Who was the chore boy? He was Gus Strand. He was Antoine Menard. He was Jack Peters. He was Terry Silvera. Over the decades, many men, young and old, did a wide range of chores from running the cream separator to gardening to serving as night watchman. We know some of their names and some of their chores. With one possible exception, none of them became millionaires, but they all had stories.

Gus Strand was the best herdsman Conrad Warren ever had. His job was to prepare Warren's registered Herefords for stock shows. His pride and joy was a team of Hereford steers he'd broken to pull a cart. He called himself "the pig man" because at other ranches where he'd worked, if a sow had too many piglets, he'd adopt the spares. He kept a copy of Kate Smith's Old Home Cookbook in his room. Pretty meager facts about a man who worked for Warren on and off for years. I didn't often mention that Gus had a fondness for alcohol which finally led Con to fire him. Is there a point at which thorough interpretation becomes invasion of privacy?

Many Chinese cooks worked for the ranch in the bunkhouse and at the "upper ranch" where the registered breeding herd was pastured in the summer. Despite the obvious origin of their names, the pay book always noted "chinaman" beside the name. Why?

MICHAND "MITCH" OXARART HARDLY FITS
THE STEREOTYPE OF A RANCH FOREMAN.

One of Kohrs' foremen, Mitch Oxarart, was from France. How does that fit in with the movie stereotype?

The classic view of the American cowboy is a slow-talking, independent type, much given to rowdy behavior at the end of a trail drive and a tendency to duck his head down and stammer "Howdy, ma'am" when confronted by a lady. Teddy "Blue" Abbott, though he came up the trail with a herd from Texas, was born in England. A fair number of adventurous farm boys were lured West by the portrayals of cowboys in the dime novels of the 1880s. There were even a couple of horse thieves.

Take Grant, Kohrs, Bielenberg and Warren out of the story, fascinating tales remain. Take away the foremen and you have the cowboys, cooks, household help, fencing crew and hay crew.

It's endless—and endlessly interesting.

Languages

"Out in the West Texas town of El Paso,

I fell in love with a Mexican girl.

Nighttime would find me in Rosa's Cantina,

Music would play and Felita would whirl."

That old Marty Robbins song, El Paso, nearly drove me crazy. An annual Border Folk Festival was held in El Paso, at Chamizal National Memorial, and I was lucky enough to be invited to participate, demonstrating forge work. My Spanish was rudimentary, at best, and since my high school teacher hadn't thought it necessary to teach us words like anvil, forge and tongs, I feared I was ill-prepared to interpret to a largely Spanish speaking audience. Meanwhile, that old song ran like a stuck record in the back of my mind.

The regular Chamizal staff, all fluently bi-lingual, politely but firmly suggested that I stick to English, but it proved unnecessary. Over the course of the event, the audience taught me all the key words. One man even watched for a while, left, and returned with a handwritten Spanish-English "dictionary" of blacksmith terms.

I was particularly heartened when they explained that just a few years earlier I would have had to use the masculine word form for blacksmith (herrero), but it was now acceptable to feminize it (herrera).

At the end of the three-day event I was surrounded by the families of several young boys who had spent hours watching me work, and they presented me with paper festival flowers which I still cherish.

No one watching me work felt it necessary to correct my grammar. They asked questions which were a combination of Spanish, sign language and timing. That is, I already knew what kind of question any particular part of my demonstration was likely to raise, regardless of the visitor's native language, and recognizing a word or a gesture was all that was needed. For example, while I'm adjusting the position of hot iron on the anvil face, I tap the anvil a couple of times. That saves energy, since as long as I continue the rhythmic hammering on the hot iron or anvil face, it bounces back up so I barely need to lift it at all. So, when a watcher gives me a curious look and imitates the back and forth bounce of the hammer, I hold the hammer handle with the tips of my fingers and let the head drop. Actually, it's easier to demonstrate the technique than to tell about it. With just a slight assist from my fingertips, the hammer continues to bounce. Then there are smiles and heads nod in understanding and the best thing is it works in any language, because it doesn't rely on language.

The song stayed in my mind for weeks after I got back to the ranch, but at least it no longer conjured up Robbins' story of a young cowboy shot down. It brought back to mind the border folk, their generosity and their smiles. It also made me more determined to learn a word or two in every language I encountered.

This ambition had actually begun a few years earlier when a Russian visitor was in the Grant-Kohrs Ranch shop. He spoke little English, but was able to tell me that my job was "cuznyetz." Then, with a teasing smile, he informed me that it was a man's word.

"How do I say 'I am woman blacksmith'?" He shook his finger at me and said sternly, "Een Siberia, ees NO woman blacksmith!"

I began to notice that merely asking foreign visitors for a word in their native languages broke down barriers. I had collected "blacksmith" in nearly 40 languages before it occurred to me that "hello" or "welcome" might be more useful.

I had learned a few by the time a small foreign tour group came into the shop. They didn't look very cheerful. I asked the frowning leader if they were comfortable with English and he gruffly replied, "I will translate for them! You don't know our language. Croatian."

Hah! I turned to the group, smiled and said "Zdravo," which means hello. Smiles broke out, and when I confessed that they have a very hard language, they nodded and began speaking to me in English! They actually knew quite a bit of the language but were apparently too reserved or embarrassed to try it in public.

This kind of incident happened over and over. A single word in Swahili (not often called for, I must admit), Japanese, Hebrew, Italian or dozens of other languages usually brought out smiles and the surprising (and somewhat embarrassing) question, "You have *heard* of our country?"

Gradually, too, I began to notice that in Europe, the word for "blacksmith" is very often a family name. The Scottish "McGowan" and Polish "Kowalski" both mean son of the blacksmith. In Asia, it is just a word for the job, and in Native American languages it is often a description of the work.

None of my language skills would get me a good grade in school. All they are good for is communicating.

BOARDS BRANDED WITH "BLACKSMITH" IN DIFFERENT LANGUAGES
WERE A SUREFIRE WAY TO CONNECT WITH FOREIGN VISITORS.

Several Ants, a Caterpillar and Other Bugs

It would have been hard to miss the caterpillar. It was bright, lime green and stood out clearly against the weathered gray-brown boardwalk in front of the ranch house.

I might have missed the ants which surrounded it, although I am usually careful to avoid stepping on the tiny creatures.

Curious, show an ant to the average six to eight-year-old and the child's natural instinct is to stomp on it. Why?

Certainly the presence of ants can be annoying, but they also break down a lot of nature's "trash." We'd be buried in decaying matter if it weren't for the efforts of various insects and molds.

The team of ants I watched had quite a system worked out. It looked as if they were fighting over their caterpillar captive, but then I realized that half were pushing it and half were pulling. At one point it looked as if some sort of conflict had begun, but they were just circling halfway around so the pushers became the pullers, and vice versa.

When you don't understand what's happening, it's easy to make wrong assumptions. Abruptly, one ant dropped out of the effort and legged it away to the south. "Hey," I said (yes, I talk to ants), "get back there!"

It didn't, but moments later another ant appeared from the north and took the place of the departed ant.

How did it do that? Do ants have tiny cell phones? Ant number one makes a call, "Hey Cynthia, I've got to pick up the kids at day care. Can you take over for me?" Ant number two replies, "Sure. You just go ahead. I'll be there in a sec."

Strength dwells in numbers. Humans have worn a path of trampled grass across the ranch house lawn, but crushed blades of grass are nothing compared to what the ants have achieved: A strip of bare dirt half an inch wide leads from a patch of antique poppies, across the lawn and down to the pasture just outside the picket fence. Close inspection reveals the annual passage of thousands of ants.

Ants are extraordinary. We had a wildfire down along the river one year, and as we worked through the night over the smoldering ground, I came upon a tall ant mound. Two and a half feet high, the top glowed in the dark like the caldera of a volcano, while all down its sides, ants were frantically carrying away the eggs of the next ant generation.

I am not a bug enthusiast. Tick fever as a small child and Lyme disease as an adult have made me very suspicious of crawly things.

Suspicious, but impressed.

The nests of paper wasps are another case in point. They hang from the undersides of equipment sheds, lurk in the lilacs and inconveniently depend from the front porch of the ranch house. Their nests, ball-shaped constructions filled with hexagonal cells, are made of chewed-up plant material.

With the exception of yellow jackets, which boil up out of holes in the ground and whose sting feels like a thump with a two by four, most wasps are pretty peaceful.

The paper wasps' policy of "live and let live" generally lasts until cool fall weather arrives. At this point they turn nasty. My personal theory (totally unbacked by scientific opinion) is that they know winter is going to do most of them in and they want to get one sting in before they're done.

I like one type of spider. It's the huge orb weaver which, with face-like markings on its back, is quite intimidating. In the blacksmith shop, however, they keep down the flies and mosquitoes and provide some interesting opportunities for interpretation. They are nicknamed cat spiders.

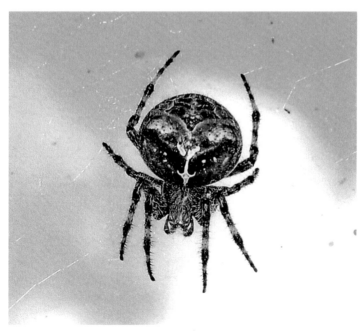

CAT SPIDER. LOOKS AREN'T EVERYTHING, AND
CAT SPIDERS WERE WELCOME IN THE SHOP.

One woman and her young daughter had been in the shop for several minutes when the mother noticed the webs. In a voice which was nearly a squeak, she confessed to being terrified of spiders. "I don't want my daughter to be afraid," she whispered, "but…." her voice trailed off.

I went to the bench and picked up a horseshoe nail ring. Then I put it on her little finger and said, "This is a magic ring. As long as you have it on, every spider in the world is a good spider!" It was a silly thing to do, but it seemed to help. I finished the forge demonstration for her daughter, while she held the ring tightly with her other hand and kept repeating, "Every spider is a good spider!" I really admired her, because if it had been ticks, I would have fled the shop myself.

A second spider incident was more amusing. A class of elementary school kids saw the webs. "EW!!!" Spiders!!" they exclaimed.

"How many of you kids like flies?" I asked. "How many like mosquitoes?" None of them did. "What do spiders eat?" I continued. "Flies and mosquitoes!" they responded. "So spiders are OK?" I asked. They agreed, although with slightly less enthusiasm.

Just then, a mosquito landed on one of the webs and instantly a large spider dashed down to it.

Almost as one, they yelled, "Yea, Mr. Spider!"

Then, making it clear to me that times have indeed changed since I was little, after a pause they yelled, "Or Mrs. Spider!"

I Go to Prison. Again.

Back in 1974, transferring from Fort Point to Alcatraz, the former Federal prison, was not my idea. Both are part of Golden Gate National Recreation Area. Fort Point is a Civil War era fort now hidden beneath an arch of the Golden Gate Bridge. My seasonal appointment there was nearly done when the Alcatraz supervisor came to the fort and told me I'd be starting at Alcatraz the following Monday.

"But I don't WANT to work at Alcatraz!" my inner voice wailed. Nonetheless, I began "doing time" there. I had been fortunate to be hired by the Park Service without a college education. I suspected that if I had any lengthy break in service, I would find it difficult to get back in.

As is generally my preference, I larded my program with all the amusing trivia I could find on the island's history. Then, at the end of one tour, a smiling visitor commented, "I never realized prison could be so amusing!"

Darn.

Starting over, I rewrote the program to more appropriately reflect the story which should be told. It was depressing.

There were compensations. I had previously done a bit of pair skating with an extremely talented skater who eventually eloped with a chorus girl from Ice Capades, and one of my Alcatraz coworkers danced with the Oakland ballet. We were down on the empty dock one day, practicing lifts and secure in the knowledge that the tour boat made so much noise we'd hear it long before it came around the point.

Unbeknownst to us, the boat had broken down and a larger, silent boat was floating towards us, roughly 150 park visitors agog at the sight of two rangers in uniform (minus the hats), one of whom had been lifted high above the head of the other. Fortunately, my partner didn't simply drop me before he darted into the break room.

Poking around the tide pools was interesting too. The prison had been on the island since 1860 and bored guards had apparently amused themselves by taking potshots at shore birds. We found everything from Civil War minie balls to .45 caliber slugs in the pools—and a "Wilkie for President" button.

When word came to headquarters that my application had been accepted at the ranch, the office on the mainland radioed the island with the news. I suspect my delighted yell could have been heard clear to the city.

It was not until I arrived at Grant-Kohrs Ranch that I realized the faintly-remembered Cinderella's castle of my childhood was actually Montana State Prison. Located at the south end of Main Street in Deer Lodge, it was surrounded by a gray stone wall with lovely turrets. As a child, I had been unaware of the significance of the bars on the windows of the impressive and surprisingly attractive brick buildings inside the wall. It was still an active prison when I arrived, though it closed a few years later.

When it became known that I held a level-one federal-law-enforcement commission and had worked at "The Rock," some bright soul came up with the suggestion that I might like to become the first female guard in Montana's men's prison. However, my mother didn't raise any fools. (I realize the common phrase is "my momma didn't raise no fools," but the grammar disproves the assertion.)

40,000 acres of ranchland are associated with the prison, dating back to 1889 when it was contracted out to Conley & McTague. In 1908,

THE MONTANA STATE PRISON.

the state had taken over its operation. This included the ranch, and the state continued raising livestock and crops and generally continuing the prison industries established by Conley.

Several miles west of town on prison land is the pistol range where I fired my semi-annual qualification courses. The armorer always made these requirements more interesting by bringing out various arcane weapons, and so I found myself firing a Tommy gun, trying to just squeeze off one shot and hold the barrel level. Alas, I had fired off a long burst and the barrel had swiveled across a shocking stretch of hillside before I could get my finger off the trigger.

One of my fonder memories was the day we fired tear gas and the wind changed, blowing it back over us. I enjoyed the sight of a bunch of apparently weeping men who had earlier been vocal about having to have a "girl" in their midst.

Now, a state prison might not seem to have much to do with the story of cattle ranching as told at Grant-Kohrs Ranch, but I found they were really tied together.

Make that "hitched" together.

Western penitentiaries, and particularly "Deer Lodge" as the prison is commonly known, produced a beautiful bit of art called hitched horsehair. Possibly originating in ancient Persia, brought to the Southwest by the Spanish, brought north by sheepherders and taken up by inmates who had a lot of time on their hands, hitching is fascinating, lovely, durable and iconic. Not all the hitchers at the prison were cowboys, but famed cowboy artist Charlie Russell told of a cowboy gone-wrong who was "hitching hair in Deer Lodge now," and the general public throughout the country understood the reference.

Both hitched and braided horsehair work is in the collection at Grant-Kohrs Ranch. Braided work is less complex, but an excellent braided bridle was made in the 1860s by Juan Rope Cordoba, a sheepherder hired by Kohrs & Bielenberg to tend a flock kept in the Big Hole Valley. Cordoba fell ill, and when Bielenberg paid for his care, he asked his erstwhile employer to bring him all the horsehair he could. The resulting bridle is a fine piece of work, incorporating fancy hair knots and red "trade cloth."

In researching the history of horsehair work, I learned a lot about the art itself and even more about the integrity of the men who practiced it and the warden who promoted it.

One of the most important things I learned was that you can't put people in a box labeled "good" or "bad."

Warden Conley was a bigot who would help a released black inmate find work. He wrote, "There is no color line in prison." He was contemptuous of do-gooders, yet wrote Henry Ford in support of one of the West's most active reformers, Elizabeth Nowell. He possessed an extraordinary ability to take actions loftier than his personal beliefs. He had been kept on as warden when the state took over. On the other hand, Governor Dixon virtuously portrayed himself as a reformer and fired Conley in 1921. However, this self-righteous governor was not above using his office to manipulate a public auction and acquire prime land on the shore of Flathead Lake when the reservation of the Salish and Kootenai was opened to white purchase.

When I was introduced to the art of hitching, Clay Mills was one of just a few inmates still practicing it. His work was famous and distinctive. One time I was in the Denver airport, wearing a belt he had made when a voice burst out behind me, "Is that a MILLS belt?!!"

I interviewed Clay in the hobby shop of the new state prison, and his status within the prison was revealed when he entered the building and everyone else fell silent, guards and inmates alike. He explained the hitching process and told me that if he did work he wasn't proud of he'd cut it up into small pieces so no one could attribute it to him. He mentioned that several inmates had asked him to teach them to hitch, but they wouldn't spend the time to do it right so he refused. He added, one kid looked like he'd take it seriously and he might teach him.

He subsequently did. The "kid" taught others and an art which seemed doomed to extinction was thoroughly revived. Today's hitchers have actually raised hitching to new levels and I think Clay would have agreed that they are willing to spend the time to do it right.

CLAY MILLS HITCHING HORSEHAIR AT COW CAMP. HE PASSED
THE ART ALONG TO A NEW GENERATION OF HITCHERS.

Feeding

If I ever open a restaurant, it won't be for people.

When it comes to people food, I'm a consumer, not a provider. But I love to feed animals: horses, cattle, chickens, wild birds, cats and dogs.

Feeding cattle is particularly rewarding. All the little stumbling blocks that can get in the way of getting work done when dealing with humans don't exist in the world of feeding cattle. You have hay. They want hay.

Once you've tossed the hay out, they either eat it or they don't. They never, however, ask you to redo the work. They never insist that you toss the hay out in a different direction and they never mumble to each other about how much better Dawn or Glen was at pitching hay.

They aren't clones though. They have individual quirks. Take Penny Rose, for example. She had that snazzy, official name when we bought her because she was a registered Longhorn. It may have been a name which linked her with her heritage, but she also had a copper-colored circle on her side that might have looked like a coin or a flower. I just called her Penny and at feeding time she was my sidekick.

Of course, she couldn't sit beside me in the truck, but she strolled along just outside. She possessed a skill which might have made her unpopular on the rodeo circuit. As it's been explained to me, the theory of "bulldogging" is that when the rider jumps off his horse onto a running bovine, if he can get the animal's head turned, it will be unable to continue running. Cattle, it is said, have to follow their heads.

Penny didn't. She could walk straight ahead while looking right in the window. This was fortunate, since her horns were so long that had she looked ahead when she was on the driver's side, I might have ended up with a pierced ear. As we rolled along, I'd tell her how terrific she was and what a good sidekick. Her facial expression wasn't revealing, but she seemed, at least, to be listening.

The routine was always the same. She'd stick by me until the last bale had been pitched off. Then she'd settle down to enjoy breakfast.

It could get a tricky if I was on the back, pitching hay instead of driving, because Penny not only accompanied me, she tried to help. With her long horns she could impale an 80-pound bale and lift it right off the truck.

Penny eventually left the herd, but for the next couple of years her place was taken by Chicken Little. As a calf, her [forelock?] stuck up

PENNY ROSE. IF I WAS ON THE PASSENGER SIDE, SHE'D MOVE TO THAT SIDE TOO.

OFFICIALLY RECORDED AS H3. THIS HEIFER WAS KNOWN AS
CHICKEN LITTLE UNTIL HER COWLICK SETTLED DOWN.

like a hen's comb and she had a startled look. She became an occasional sidekick, but as she got older the startled look faded, the growth of horns diminished the chicken-comb look and she finally faded back into the rest of the herd.

Once the last bale was pitched off, whoever had been on the back of the truck would hop down and climb in the cab to thaw out as we made a wide loop and drove slowly back alongside the herd, counting to make sure everyone was there and watching closely for any signs of illness, injury or impending motherhood. That was when it was really important to not just look, but to think about what we were seeing. It also acquainted us with individual cows. We got to know which ones always stuck together. We learned which cow was at the bottom of the pecking order and which ones were convinced the next cow had better hay.

A lot of theories of feeding exist. Traditionally, ranchers got up in the early morning dark to feed and grain the horses which would be pulling the hay wagon. The cattle got fed early and once all the feeding and milking chores were done, the ranchers ate.

Times change: Statistics clearly indicate that feeding pregnant cows in the late afternoon results in more calves being born in daylight hours. The reason put forth is that cows spend a couple of hours eating and

FEEDING THE COWS.

121

re-eating (the details aren't particularly tasteful unless you're a cow) and over several more hours the feed makes its way through their complex digestive systems. Then, this all-important process complete, the cows are ready to calve in the morning.

Technology has brought about significant changes. Webcams strategically placed keep some ranchers informed on the status of the herd. In some outfits, any cow which has had problems calving is automatically sold. Even the ritual of tossing out flakes of hay is changing as machinery scatters the hay at some ranches and at others the bale strings are simply cut and the cows pull the bale apart on their own.

All of this is good business, no doubt, but I'd miss that long slow drive. Sometimes there are too many animals to get to know them all, but there will always be a few which stand out. And like the 4-H and FFA animals and my bottle calves, the ones you get to know tell you a lot about the ones you don't know.

Do You Speak Agriculture?

Whatever you do in ranch country, don't suggest going out to rope some "doggies."

A doggy is a canine. A "dogie" is an orphan calf.

It's an odd word, and peculiar to the West. Its origin is frequently described as "unknown." It is also speculated that the word evolved from the pot-bellied appearance of calves which were subsisting on grasses in the absence of a mother's milk.

That's close, but I think the definition provided by D. J. O'Malley might be even better.

O'Malley could fairly be called the original cowboy poet. Born in Texas in 1868 and transplanted to Montana, he wrote poetry which remains current today, notably a sentimental classic, *When the Work's All Done This Fall.*

He wrote stories of those days too and mentioned asking a Texas drover why the "orphan" calves were called "dogies."

They weren't actually orphans, was the reply. Farmers raising milk cows would take steer calves away from their mothers and feed them the less valuable skim milk while selling the rich milk and cream. This gave the calves a pot-bellied appearance. Since steer calves were not needed in a dairy herd, the farmers would happily sell them to trail outfits passing by. The cowboys called these calves "dough bellies," and in time this evolved into "dogies."

That's a word which has a comparatively recent history. The most ancient word I've run across which is still in use is "Bossy," as a common name for a cow. The Latin word for cow is "*Bos*."

It's interesting the number of phrases which are still in use which have to do with agriculture. Even in the 21st century, "Strike while the iron is hot," is still in use, and many of the high school students I work with are familiar with the phrase. They know what it means, but they don't know what it meant.

The best temperature for working hot iron is between 1400 and 1800 degrees Fahrenheit. Colder iron is harder to work, and its molecular structure can be damaged by working it at too low a temperature.

Take a wire and bend it back and forth until it breaks. That's called work-hardening. The same thing happens to iron which is hammered below a good working temperature. There are times when work-hardening is desirable, but that's getting rather far afield. "Afield?" there's another one!

Another phrase still in use is "Too many irons in the fire." As a blacksmith, I know that if you are heating too many pieces of iron at once, it can be hard to keep track of the heat on each of them. You are likely to glance back at the forge just in time to see white sparks flying up out of the coals. This means you have just burned the carbon out of the metal and ruined some piece of work. Perhaps that's the origin of the phrase. Perhaps not.

The hand pump I put in the shop just recirculated water in a tank. It was there to teach kids, not to provide fresh water. If it wasn't used for a couple of days, a leather gasket dried out and the pump wouldn't hold water. Then it was necessary to pour water into the top to "prime the pump."

"Make hay while the sun shines." Why? Because as mentioned earlier, hay put up wet can mold and even generate enough heat to start a fire. Moldy hay can sicken horses. Admittedly, older cows often seem to enjoy brown, moldy hay, though even those cows will snub white mold.

The only advantage to moldy hay bales is on a sub-zero morning when your hands are freezing. You cut the bale strings and—with considerable effort—pull the bale apart. A puff of white rises up and you slide your hands down into the warmth generated by the mold. Bliss…ACHOO!!

Most kids have at least attempted to do a cartwheel, at an age when falling over is only embarrassing and not painful. Despite the number of wagons at the ranch, it wasn't until I contemplated making a repair to a two-wheeled cart that the word made a connection between the spokes of the wheel and the whirl of arms and legs. "Oh! Cartwheel!"

Undoubtedly books would provide long lists of these sorts of words and phrases, but I'd rather discover them personally, one by one. An "aha!" moment is much more satisfactory.

Cool, Clear Water. Or Any Water

In one of my favorite books when I was a youngster, the villain stole water from a neighbor. At one point, the hostilities included the heroine holding the evil rancher's foreman at gunpoint. That was my introduction to the concept of water rights, and though I couldn't have said I fully understood all the legal aspects, I can now confidently state that I will NEVER understand all the legal aspects!

I'm in good company. Water law in the West is unbelievably complex.

East of the 100th meridian, a lot of folks can't imagine that people could have an entire river running through the middle of their property and not be allowed to use it to raise crops, water livestock or build a gigantic water park complete with fountains and waterslides.

The central theme of the ranch is open range grazing, but without some understanding of the availability of water (or lack of it) visitors won't fully understand not only how, but why some ranches cover thousands of acres. Sparse vegetation in unirrigated desert land means cattle have to stay on the move to graze.

I found it easy to come up with dry facts about water, but surprisingly hard to make the story come to life. It took a while to realize why: I never did any irrigating.

I've built fences, pulled calves, driven the baler, hammered hot iron, fed chickens, and a dozen other jobs. Some jobs I did only once, some were as regular as the sunrise or the seasons, but irrigating? Never.

There are 25 water rights on the ranch. Two of them date back to the 1860s and Johnny Grant. In the simplest possible terms, a lot of water rights in the West fall under the doctrine of "first in time, first in right." That is, the first person to use a water source owns the right to use that amount of that water. The difficulty comes when a person upstream diverts the water before it reaches the person with the right. The timing and amount of water used to irrigate crops is critical, so it is, perhaps, not surprising that in a short water year, ranchers can't always wait for due process to resolve disputes. I thought I might have been transported back in time a couple of dry decades ago when our irrigator resorted to a threateningly raised shovel to persuade an upstream land owner to give us our full water right.

My own awareness of the problem came when I noticed our cows' hollow flanks. I hurried through dense brush to the stream which should have been providing them water and discovered that someone upstream had diverted every drop of the flow to create a huge ornamental pond. I opened a gate and let the cows and calves into the next pasture where there was water. They ran—bellowing and tails flying—to the stream, where they drank and drank and drank. If I'd had a shovel at that moment, I might have raised it myself.

Snowpack and rainfall aren't a mystery. It is a constant and serious point of discussion in this part of the country, but I rarely actually *work* with snowpack and rainfall. It's something I watch with anxiety or hope. A slim glimmer of understanding occurred a few years ago when drought decreed that we did not have water to waste on yards. That was fine, as far as a mere lawn went, but fruits and vegetables need water and can't wait for restrictions to be lifted. That's food. When an occasional drizzle of rain came at night, I put a barrel beneath the downspout on the rain gutter, hoping I'd collect enough to keep a couple of potato plants and peas alive. It was amazing. A minute amount of precipitation across a roof produced a full barrel of fresh water by morning. The plants loved it! I've continued to collect water from the downspouts ever since. It has the added advantage of allowing me to get enough water to thirsty species without drowning nearby plants which don't require as much.

So, I understand the importance, I know some of the hazards, I know some of the solutions, but I still can't convey the hard and constant work of irrigating.

Coworkers whose daily job it is to move sprinkler pipe, maintain ditches, check headgates, and slog in heavy rubber boots through hot and mosquito-infested pastures have an understanding I lack. I can explain how they start in the earliest light of morning so the heaviest burden of the work is accomplished before the heat of the day. I can tell of their pride in getting the water evenly over fields which are far from level, and greater pride when all that hard work results in an excellent hay crop.

A CURIOUS COTTONTAIL AT THE HAYSTACK.

All of which is truth and history, but it's second-hand, and the visitors receive it third hand.

Years ago, Fred told me how his father would go out into the flood-irrigated hayfields and when he found a high spot that wasn't getting enough water, he'd work there with a shovel to throw water from the lower area across it. He'd do that for hours. And when haying time came, every inch of the hay on those high spots was as green and tall as the rest.

That's quite an accomplishment. I don't really wish to do it, but I wish I'd done it.

Directions

Interstate-90 covers over 500 miles within the boundary of Montana, and anyone who has driven it may be forgiven for thinking they've seen a lot of the state. Generally, it crosses east to west (or, naturally, west to east) but where I-90 passes by the ranch, it takes a north-south turn which leaves many out-of-staters geographically challenged.

The Deer Lodge valley is about 30 miles long and seven miles wide. The hills and peaks which surround it seem very close and it's easy to get the impression you can see it all from the freeway. Take a side road east or west, however, and brand-new country is revealed over every rise. There are little enclaves which might have been the ruins of an abandoned town or simply an accretion of houses as youngsters grew up and built their own homes on a family homestead which has since been abandoned.

A ranch in the southwest end of the valley raised sheep, and I was invited to a shearing. The directions were typical: Go south on the frontage road and turn right just past the pasture with the black baldy heifers. Well, that seemed plain enough. I knew black baldies were white-faced black cattle, and heifers were young females.

What I didn't know was that the ranch in question had just sold the heifers and they were no longer in the pasture. I blithely passed by the turnoff and rolled south until the feeling that I had gone too far could no longer be ignored. I saw a lot of new country before I flagged someone down and got turned around.

Country directions are tricky. You'll be told to turn at the old Johnson place. Watch out! "Old" is a clue. It probably means that the place in question no longer belongs to Johnson, but will forever be identified that way by long-term residents.

Newer arrivals may not be aware of this and when asked for directions will say, "Sorry, I don't know any Johnsons around here…"

The ranch encompasses 1600 acres, and even within that small an area it's possible to misunderstand directions. I learned to be irritatingly specific when told to "close the gate" or "see that the heifers have water."

Which gate? "The red one." Which red gate? "The one on Cattle Drive." Which red gate on Cattle Drive?!!! There are nine red gates on Cattle Drive and more than 250 gates on the ranch.

Which heifers? "The yearlings." Where are they? "We just moved them." Where? "The pens." Which pens?!!! There are yearlings and two-year old heifers and although "the pens" are most likely the ones behind the red barn, there are other pens and, for that matter, there are also several red barns.

Like the confusion which arises from "the old Johnson place," we've also had name changes in the last 40 years. I learned the names of the pastures from Con Warren. Humber was named for an early homesteader. It is now called Westside 1-4. Our southernmost field was called the Cemetery Field because it was across the road from the Deer Lodge cemetery. Taylor Creek runs through that field and its name has somehow changed to Taylor Field. The Stuart Field (named for another early settler, Tom Stuart) has been divided in two by the NPS and is now the Stuart Field and the Stuart Annex.

Name changes can be confusing, but when I contemplate starting a campaign to keep the old place names, I recall some research I did on stupid Montana laws. A session of the Montana legislature in 1911 sent a memorial to the U.S. Congress requesting that there be no changes in old place names. Had they been granted this request, and had it been adhered to in the strictest possible sense, the road signs at bridges over the Yellowstone River might show the Cheyenne name for the river; mo'ehe ohe'e. And good luck with that!

Hunting

I spent a lot of time hunting on the ranch.

No license was necessary for these hunts, the quarry usually being an escaped cow, horse, flock of chickens, porch pillow or flower pot.

Escaped cows do tend to wander, particularly when they've grazed down the best grasses in their pasture. Luckily, since they are herd animals, they are likely to stay in a group, and it is easier to locate a bunch of 20 cows than just one. At first I really worried when a calf got

out by itself, but I learned it would generally stay along the fenceline. Then it would get hungry or its mother would bellow and it would slip back in through whichever wide spot it had slipped out through.

A rope hangs in the back window of my truck. These days I mainly use it so I can spot my white truck among the hundreds of other white trucks in any large parking lot. However, I have needed it on occasion, and the thing about a rope is that when you need one, it's too late to go find one.

One early morning call from the Sheriff's office informed me that cattle were out and headed towards the freeway. I found them, but they weren't from Grant-Kohrs. They were from the nearby fairground and were knee-deep in fresh grass. Nonetheless, I started to try to haze them back to safety. If a steer could say, "You wish!" that was the look they were giving me. However, these were obviously roping steers—animals used in roping competitions. So, I hauled out my long-disused rope, shook out a loop in it and said, quietly, "You know what this is, I expect." Luckily none of them knew I couldn't have roped a fencepost from 10 feet away. They bunched up and headed back on a trot. Alan arrived about then and the two of us soon had them penned up again.

Horses are more of a problem. Theoretically "herd animals," horses are not reluctant to strike out on their own in search of greener pastures.

Pastures, by the way, really are often greener on the other side of the fence. That may be due to irrigation, wiser grazing, fertilizers or just the illusion of denser growth seen when looked at from a lower angle. It's important to keep horses from eating too much rich, spring growth; it can lame them, sometimes permanently. It's called foundering, and one of our horses showed remarkable sense when the swelling in his feet became painful. I came out one morning to find him standing with all four feet in a large, deep water trough.

Of course, with horses, the easiest way to catch them is just to rattle a can of grain, and they'll come right away. I have never yet seen a horse turn its nose up at a can of oats.

With escaped chickens, it was a matter of timing. The easiest way to recapture them was by simply waiting until evening feeding time and calling "Hey you turkeys! Get back in here!" Don't worry about calling them turkeys. They don't know they aren't turkeys.

Occasionally I had to resort to my chicken imitation. A properly phrased, "Burawwwk-buk-buk" generally drew them home from the nearby vegetable garden or bug-infested tall grass. I don't know whether I was using the right chicken words. Perhaps they decided "Burawwwk-buk-buk" was just "human" for "come and get it or I'll give it to the chickens next door."

In fact, their eggs were often harder to find than the hens themselves. It's called "stealing a nest," which is hardly fair to maternally inclined hens. They weren't making off with another hen's nest. They were stealing away to build a nest which wouldn't be raided by humans. They were pretty crafty. For a while, they tucked the eggs behind the wide-open door to the coop. It was very rare for us to shut the door while we were on the inside, so this bit of sneakiness wasn't discovered until an enormous aggregation (or would that be "eggregation?") of eggs had been amassed.

Some hens started caching eggs under the ramp into the coop. It wasn't just one hen—there were brown and white eggs and little banty eggs. That spot was doomed to failure, however, since the hens would be shut into the coop at night and unable to provide the necessary "egg-sitter."

The runaway porch pillows were a real mystery. I'd lost more than one. They were small, just the right size for a cat to curl up on when the sun hit that side of the porch. Having run out of small pillows, I put out a larger one and the mystery was finally solved. A hitherto unsuspected raccoon colony under the trailer had been dragging the small pillows through a hole under the trailer skirting. The larger pillow got stuck. Busted!

The wandering flower pot took longer to track down. It would vanish and I'd generally stumble upon it while the cats and I were on one of our evening walks; out in the horse pasture, behind the woodpile or down by the creek. The culprit, spotted at last but never apprehended was a very large, gray-striped feral tomcat. The motive was simple: The pot contained a large and healthy catnip plant.

Archeology of Sorts

Sometimes, driving across the state or hiking through the woods, I'd catch a glimpse of purple or orange. I learned that closer inspection would inevitably lead to an abandoned home. The purple would be a lilac in bloom. The orange would be a poppy, and as I approached, I'd see the huge leaves of rhubarb.

There'd be no other sign of previous habitation. Sometimes, the logs and boards had long since rotted away. Sometimes they had been hauled away and used in another hopeful spot.

The lilacs, poppies and rhubarb stayed behind, however, dying back in the winter and returning optimistically in the spring. They never gave up hope.

I might not have understood the implications of these small gardens if I hadn't worked at the ranch.

Having prospered, the Kohrs hadn't abandoned their home. A long line of lilacs forms a hedge between the house and the garden. A row of orange poppies grows outside the greenhouse. It dies back completely in the fall, and its ragged, shaggy leaves look like some sort of thistle as they first appear in the spring. Down in the flower garden, a patch of rhubarb grows every year, apparently disregarding the fact that it is not in the least ornamental. It does make a great hiding place for barn cats who delight in leaping out to startle you as you pass by.

Plants aren't the only "relics" telling the history of bygone days.

Barbed wire fences tend to stretch and sag with time. To tighten them up, ranchers often added an iron bar with a hook at either end to pull sagging wires together. It wasn't until I started restoring a potato planter and researched the pioneer history of raising "spuds" that I discovered the iron bar was part of a long conveyer called a "chain" on a potato harvester.

The mystery to me was why so many fences all around the county had recycled potato chains woven into them. Eventually, I learned the Deer Lodge Valley had been a potato "capital" of sorts, providing seed potatoes to much of Idaho and Washington. I had to revise my mental picture of the valley as a place for grazing, haying, grains and livestock.

A former ranch hand told me of the 25-acre potato field on the west side of the river. That, in itself, was a surprising news flash. However, more to the story revealed that each spring, when he plowed that plot, rainfall would wash the dirt off hundreds of flint chips on the surface, which would catch the light as sunshine returned. We had known that many tribes passed through the valley on their way to hunting grounds. However, we had pictured their passage as a fleeting thing, not a lengthy encampment to make the arrowheads and other stone tools of the hunt.

Other archeological finds were made within the ranch boundaries. I came to think of one such as a sort of "Brigadoon," an enchanted land which only appears for 24 hours every hundred years.

My Brigadoon only appears for a few days each year, and even then only if it's a hard winter. The rest of the year it's off-limits to all except muskrats, other small mammals, bugs and birds. A pair of great-horned owls nested there and walking nearby one time I scared up a golden eagle. Tracks of deer and fox identified it as good habitat.

Though it does support the growth of alder, willow and cottonwoods, it's less like a ground than a floating island of mint, forget-me-nots and death camas. I'd think I was stepping across a narrow creek onto a solid green bank and—splash! I'd be up to my knees in thick mud and icy water and everything above the water line was swarming with mosquitos. I avoided the area after that.

But once a year, in the dead of winter, it allows people to wander through it.

I might never have rediscovered it, except that one sub-zero day I let Beechnut pick the route for our walk. The ground was frozen, although water still tricked through some of the deepest channels.

The attraction for the barn cats was the variety of mice and other rodents who make their homes in the tangled underbrush, and even in summer it had seemed to be a favorite hangout for them.

I wasn't the only human to have explored the place. Remnants of children's forts, built of nails and old boards, or just string and dead branches remained. Some of the forts seemed to be of recent vintage, and some probably dated back to the grandparents and great-grandparents of today's builders. Faint scraps of kites tangled in the upper branches of the cottonwoods spoke of a carefree time before drones.

One woody shrub caught my eye. With the ground frozen, the cows had been able to get to it, and had rubbed and nibbled it down nearly to the ground. The cats made quite a ritual of rubbing against it, and I sniffed at a bit of peeled bark and identified it as honeysuckle.

A subsequent exploration later in the year revealed a small lilac but no sign of previous habitation beyond the children's constructions.

It seems there is no end to the things I don't know, and that's a good thing: The day I stop learning, everything for the rest of my life will just be repetition.

350 Ground Squirrels or .66 Bison

One of the hardest things to explain to ranch visitors from Europe and states east of the 100th meridian is the acreage required to be a "Cattle Baron."

Very few people can picture an acre, let alone the hundreds of thousands of acres of open range used by western cattlemen in the late 1800s and early 1900s.

For comparison's sake, at least until recent years, a typical house lot was about a quarter of an acre, so four lots comprised about an acre. Back East, with more rainfall and an often-milder climate, an acre might support a couple of cows. Out West, in unforested land which receives little rain, where the soil can be sandy and alkali-laden, where temperatures are often high and humidity low, it can take over 100 acres to support one cow with a calf. In fact, there are a few areas which require as much as 250 acres. In that extreme case, someone claiming to have a 1,000-acre ranch might just be raising four cow-calf pairs.

An AUM (Animal Unit Month) is a way of calculating the carrying capacity of a particular area. With runoff from the nearby mountains to keep the valley green, the Deer Lodge Valley can support animals on less acreage than those marginal ranges. To put it as simply as possible, the calculation on the valley floor is roughly eight acres for either one cow with a calf, or one horse, or five sheep, or 350 ground squirrels, or .66 bison. The drier, nearby hills require 36 acres.

A horse requires more feed than a cow because it doesn't digest it as efficiently. That's why you can step on fresh horse poop with relative impunity, whereas it is highly recommended that you avoid stepping in a fresh, squishy cow pie. Dried, a cow pie makes a fair frisbee for a chip-chucking contest, while buffalo chips aren't as flat, though they do hold together well when you toss them.

More sheep can be grazed on that acreage because they eat less. The same is true of ground squirrels, although, while I hear there are people who deliberately raise sheep, anyone raising ground squirrels is likely to be run out of the country.

As for bison, they take more land than cattle, although like cows they have four stomachs to process the grasses. However, their digestion is not as thorough as a cow's. I've never actually seen .66 of a bison, but I should think it would find it hard to get around, even in the best grazing conditions.

None of my agriculture reference books give the AUM equivalent for grasshoppers but one extension office publication warns that even light infestations of six or seven grasshoppers per square yard in a 10-acre hay field will eat as much hay as a cow,

Then there's the weather. Rain, snow and temperatures vary widely from year to year and decade to decade, but 10 inches of precipitation annually (counting melted snow) used to be the standard calculation for the valley. Combine that with the fact that every drop of water in the West is legally owned by holders of water rights, and even some otherwise good grazing land is not useful if livestock can't get to water.

The only exception to "adjudicated" water rights is runoff from a roof, but once it hits the ground, it belongs to someone. Usually someone else.

Rustlers and predators are still a factor, though "virtual" theft is now more profitable than the old time masked bad man. For those with a lawless mentality, computer chips are more profitable than cow chips.

Games Critters Play

I am often accused of anthropomorphism. I'd plead guilty, except I don't think I'm attributing human characteristics to animals. I just believe animals have their own senses of humor and compassion.

Take the raven and the hawk. When a little bird chases a bigger bird, it is called "mobbing." It's a peculiar word, because while it implies a flock of small birds, it is often just one. Anyway, one day I watched a crow mob a hawk. The aerial chase went on for some time, with dives and agile twists. It would seem to be over for a moment and then the crow would return to the chase.

Abruptly, the hawk managed a midair flip which put it in the position of chaser and the crow flapped like mad to escape. I have no doubt the hawk could have caught it, but that isn't how the drama ended. They raced toward the cottonwood trees and I was hoping the crow would manage to escape into the brush.

That wasn't its intent, apparently. It made one final swoop and landed on a cottonwood branch. The hawk landed on the same branch, about three feet away, and the two sat there, catching their breath and not even looking at each other. If I were really anthropomorphic I'd have imagined one saying to the other, "That was pretty good. Do you want to try for best two out of three?"

Then there was the calf and the fox. I was out in the calving pasture (that being where we put the cows when they are due to calve) when I noticed a fox facing a young calf. It would take a tentative step towards the calf, which would jump back. Then the calf would step towards the fox, which would jump back. I wasn't concerned. For one thing, a fox is not going to try to take down a healthy calf, and for another, the calf's mother was standing by, watchful but unworried.

This had been going on for several minutes when a sharp, wheezing bark caught my attention. Unnoticed by me, another fox had been sitting on the sidelines, watching the game. The first fox took the hint and the pair trotted off, the cow turned back to grazing and the calf turned back to its mother.

Calves have a lot of fun. I often went out in the early evening just to watch them. Two or three would be running around together, and gradually they'd gather up the rest of the calves, often to the cows' annoyance. A cow would have just gotten her calf mothered up to her when a rowdy bunch would race past and the calf joined them. They'd tear around the field for a while, ignoring the cows' irritated moos before dropping out, one by one, until quiet reigned. Another game, apparently enjoyed by cows, steers and calves alike is "King of the Manure Pile." (Or, in the case of cows, I suppose that would be "Queen of the Manure Pile.")

Peanuts was a cow pony. He lived to work with cattle. One morning I heard his high-pitched whinny and felt my trailer tremble a little—not from an earthquake but from stampeding livestock. I hurried out and found someone had put Peanuts in with the cows and calves, and he was having a high old time. He'd run them all down to the northwest corner of the large pasture, catch his breath and run them back to the southeast corner. He might have been having fun, but the cows weren't. Luckily, he was easy to catch, and we spoiled his "sport," although the cows had probably found it unsporting.

Bottle calves are great for public contact. I would be surprised to learn that any child who got to give a bottle to one of our "orphans" forgot it any time soon. Most of these calves weren't actually orphans. Cows often reject one of a pair of twins. Raised on a bottle, the calves were generally very tame, but there was one standout: Wally, the Jogging Calf. Normally you can halter-break a calf to stroll along with you, and they will occasionally run a few steps before coming to a halt, but Wally hit the ground running. He enjoyed it so much that I enjoyed taking him with me on my morning runs along the frontage road. He stuck close to my side. With a bright red bandanna around his neck, he was often assumed at first glance to be a well-trained dog. That the drivers of passing cars soon realized their error was obvious as brakes were abruptly applied and we received unbelieving looks—and laughs.

Luckily, I had seen enough military dramas to be able to watch a pair of sandhill cranes close up. They were in the south pasture and apparently love was in season. They jumped up and down as if they were on trampolines, bowed to each other and seemed to be throwing something. It wasn't too far from my home, but I wanted an even closer look. I crept out and started crawling towards them, trying to keep as low a profile as possible, in the best style of soldiers making their way through concertina wire with shots whistling past overhead. They didn't spot me, and I got close enough to see one toss a stick into the air. Unusually large snowflakes began falling and soon the scene looked like a classic Japanese print. I slithered homewards in the snow, finally standing up when I figured I wouldn't spook them. Perhaps their dance wasn't strictly a game, but it was well-played.

Jerks

In 1989, Montana celebrated 100 years of statehood. Such anniversaries are traditionally times to brag about the courage, brilliance, integrity and all-around wonderfulness of our forefathers.

The Historical Society in Helena was bombarded that year with questions from folks doing research on their pioneer ancestors. Dave Walter, who introduced me to the art of historical research, wondered why all these ancestors seemed to have been governors, captains of industry, valiant warriors and brilliant innovators. Surely, he thought, someone's great-grandfather must have been a failure, a fool or a crook.

And so, Speaking *Ill of the Dead* (subtitled "Jerks in Montana History") was born. Dave lured me into the program and I began to live a double life, outside the confines of the open range cattle era and deep in the archives of the Society.

At the state history conference in Helena that year, a program we had feared might be greeted with outrage turned out to be one of the hits of the three-day event. My brief portion of the program was about the 1889 Constitutional Convention, in particular the debate about suffrage.

These days, suffrage is automatically thought of as a woman's issue, but back then there were delegates who became highly selective and remarkably offensive in their views on voting rights for men. One delegate proposed giving the vote to naturalized Germans and Irish, but denying it to Italians. One of my favorite arguments against votes for women was that it would put women in contact with the contamination of politics. A refreshingly candid observation.

Over the next several years, historians tackled such reprehensible characters as "slob hunter" Sir St. George Gore, the KKK, "The Reverend Leonard Christler and his Nine Commandments" and Mary Gleim; "Missoula's Murderous Madam." My personal preference was to take on groups rather than individuals. I particularly delighted in finding quotes which revealed extraordinary stupidity, cupidity and arrogance.

For example, during WWII, the Anaconda Company was desperate for workers, but continued to fight hiring women. In the publication "Women of the Washoe," author Bob Vine wrote, "By then, the workforce included 'a number of workers...without one hand or one arm, some without a leg or a foot, two who were blind in one eye. There were seventy-three handicapped workers, 363 over sixty years of age, 92 over seventy and four over eighty. Eighty-seven women were employed, 78 in the office and accounting department, six in the laboratory, and.... dramatic pause....three women in the foundry.'"

My program on smelter pollution included this anecdote: Copper King Clark maintained that "...the ladies are very fond of this smoky city [Butte]...because there is just enough arsenic...to give them a beautiful complexion." The Anaconda Company kept records of its emissions from the smelter in Anaconda. Between 1914 and 1918, they recorded an average of 75 tons of arsenic thrown into the atmosphere every day. The smoke drifted downwind and poisoned the Deer Lodge Valley. Spills from settling ponds poisoned the Clark Fork River."

Popular as the program was, (and with plenty of candidates for future ridicule), those of us who gave the programs began to weary of the game. Finally, I proposed a sort of faceoff: One of my heroes against Jon Axline's chosen jerks. Jon was the historian for the Department of Transportation. Judge Hiram Knowles was everything one could wish for in an ancestor. He was my hero. William F. Meyer was everything one would hold in contempt. He was Jon's jerk.

It is amazing how hard it is to make honesty sound interesting, while the slightest peccadillo can be made exciting or at least amusing. Though the audience was courteous, it was clear that Knowles didn't stand a chance against Meyer.

It's a melancholy fact: People find jerks more entertaining than heroes.

Two elements of my jobs in the Park Service don't fit well into the foregoing tale, but before this story draws to a close, I'll add them because—as with chocolate chip cookies—even seemingly minor ingredients are essential parts of the whole. Besides, what would a ranger story be without a rescue or a drawn pistol?

I Draw My Gun

To quote the Lone Ranger, "Return with us now to those thrilling days of yesteryear."

In the early 1970s, few people had ever seen a female ranger in uniform. The first speeder I pulled over (65mph in a 35mph zone) smirked as I approached his window and said, "Oh, it's a little rangerette! Are you going to give me a ticket?" His wife said, "Well, that was stupid!" a moment before I said, "May I see your license?" He deflated, gratifyingly.

At one point, the Yosemite superintendent contemplated having female rangers wear the 2" Chief, because a 4" barrel "looked too big" on us. My duty weapon had a 4" barrel, and I suggested with unnecessary perkiness that I could have it anodized pink and carry it in a black lace holster to preserve my femininity. He gave me a look which would have soured milk, turned away and, as far as I can recall, never spoke to me again.

Five years passed between the day I started doing road patrol in Yosemite and the day, three parks later, I left Grant-Kohrs Ranch for the Federal Law Enforcement Training Center in Georgia to finally qualify for my commission.

Early on at Grant-Kohrs, I was expected to wear my duty belt (gun, mace, handcuffs, speed loaders) any time I was on duty. This included while giving tours of the ranch house. A visitor quite naturally asked why I was wearing it and I told him it was in case he stepped off the carpet runners. The group and I had, by that time, established a pretty relaxed relationship, so (luckily) he laughed with all the others.

Apart from escaping livestock, the phantom poacher and the occasional cow-chasing dog, the biggest problem I faced was burglar alarms.

Not burglars. Just alarms. The local police were very good about meeting me for countless false alarms, and treated every callout as if they expected to find an intruder. Even during the shakedown period with the first intrusion system (which featured 64 nighttime false alarms in one month!) they never grouched.

During some searches, such as when there had been an escape from the nearby State Prison, I entered buildings with the gun drawn, particularly if I was working almost alone. I say "almost" because the calico cat, Victoria, generally accompanied me on these expeditions.

Alarm responses at night were "call backs," as I was off duty then. My favorite callback was the second alarm one night. A detector in the formal parlor had gone off earlier, and I had a feeling it was going to go off again. So, instead of going back to bed, I nodded off in a chair, still wearing all the defensive equipment. Sure enough, the siren woke me again, and I went straight to the parlor.

It wasn't a false alarm. Not exactly. It had been triggered by a mouse.

Now, a tiny mouse would not ordinarily have set off an alarm, but the mouse in question had climbed up a lace curtain and was perched on the curtain rod in front of the motion sensor. It escaped, which was just as well, since my handcuffs would never have restrained its tiny paws.

In my decidedly tame law enforcement career, I only came close to pulling the trigger twice—and each time would have been a fiasco.

Another ranger was on duty one evening when the ranch house alarm went off. I called on the radio and received no response. I began the search, but there are twenty-three rooms in the house, not counting closets. By the time I reached the second floor landing I was getting pretty concerned, and my gun was in my hand.

The missing ranger was blonde, and as I peered over the landing, I saw a blonde figure lying on the kitchen floor. I nearly convulsively pulled the trigger, but saw just in time that a recording box was protruding from the figure's side: It was a "Resusci-Annie" used for teaching CPR.

139

The absent ranger was fine, the alarm was false and no embarrassing hole had been blasted in the ceiling.

Some years later a new fire and security system was being installed and all the furnishings from the ranch house had just been moved up to an empty building which had been a Catholic school. I hadn't had time to acquaint myself with the school or its intrusion system when one of the seemingly inevitable alarms went off.

I searched, room by room, and was using the approved method of standing to the side while pushing a door open. A cautious look revealed a dark landing leading to three or four steps down. There, crouched below, were two people, their hands apparently up in a defensive posture. I fumbled for the light switch and found….two nearly life-size crèche figures! Again, a tightened finger might have given my NPS friends something to rib me about for the rest of my career.

Rescue Ranger

Far more often than law enforcement encounters, rangers are called upon to rescue stranded, injured or lost park visitors. Our patrol cars in Yosemite doubled as ambulances, and instead of the highly-trained EMTs of today, we were only required to have advanced first aid cards.

A couple of incidents come to mind, neither one of which would have earned me the Park Service's coveted Harry Yount award for excellence. However, they were certainly learning experiences.

The first was a simple carryout of a visitor with a sprained ankle. The other rangers and I took turns carrying the litter, and while it was not dangerous, it was hard. My arms ached. My shoulders ached, and I wanted to stop and rest before I dropped my end. It gave literal meaning to "holding up your end."

I didn't drop it. Not because I found some mysterious reserve of strength, but because the simple fact was it was my job. I'm not sure I ever had ever taken on more than I thought I could handle before and what I learned was that I could handle more than I thought.

What I learned from the second rescue was a sort of delayed reaction lesson. My patrol partner and I heard shouts for help from a couple of kids who had ventured too high on a steep bit of tumbled rock near the base of Royal Arches, a granite formation on the north side of Yosemite Valley. It was getting dark. My partner was a proficient rock climber. I was not, despite having been manager of the Yosemite Climbing

Shop at the time of the infamous "Yosemite Riot of July 4, 1970." We scrambled up the scree until we were within range of the boys. Then I sat, braced against an oak tree and played out climbing rope as Butch traversed across.

Under his instruction, the kids made their way across and down.

It was only later that I cogitated on the event. The rope would have been no use at all to protect him if he had fallen, so he probably intended it as nothing more than a guide in the dark. Well, that made sense, but another possibility existed. Maybe he had just left me by the oak tree (Quercus chrysolepis) to keep me out of his hair while he dealt with the situation.

If I ever run into him again…I won't ask.

Now What?

Every day on the ranch has had something to offer, and it will continue to do so, not just for me but for everyone who experiences it. Stories of the open range days of the late 1800s will continue to reward diligent researchers, even as today's events will become tomorrow's history. And the plants and animals, the land and the turning of the seasons will write their own tales.

The story isn't over, but this book is done.

ZORRO, A YOUNG PORCUPINE, GENEROUSLY DONATED A FEW QUILLS TO ME.

141

Mrs. Schneider's Legacy

More than half a century has passed since my English teacher assigned the writing of the class poem to me. I have written a lot of doggerel since then—short, silly poems composed quickly and as quickly forgotten.

Two which I did not toss out remained in my cluttered desk. Neither could rightly be called "poetry."

I kept one because it reminded me that ranching has a special language which would have perplexed me when I came to Grant-Kohrs Ranch. I realized that I wouldn't have understood it initially and tried to keep this in mind when giving programs to visitors. It was about loading cattle into a big stock truck.

I wrote the second poem when I had to move to town. After 27 years on the ranch it was not easy. My attempt to express my feeling was as sentimental and cliché-ridden as my long-ago class poem, but running across it years later, it was still true. Although I had been fortunate enough to call the ranch home for many years, it belonged equally to the thousands of people who visit it every year. It would be wonderful if I could help them have at least a little of that sense of ownership.

Brace yourself. Here they are:

Counting Cows

Can I count up to forty? I thought I could until,
We separated pairs one day with one pot load to fill.
We ran them down an alley, by twos and threes and eights,
While Dave and Dave were waiting there beside two swinging gates.

Bred cows to the left now; calves go to the right,
Dry cows back and through a gap that's nearly out of sight.
Only count the bred cows, Louise will count the calves,
Don't count 'em till they're fully past, don't try to count by halves.

Was that a yearling heifer? I thought she was a cow!
They'll have to get her out of there. She's with the wrong lot now.
And while they're cutting her back out, another little band
Runs by, but, Wait! That brockle-face has got our neighbor's brand.

Oh no! The drys have found themselves a loose and rotten board.
They're breaking through the hole right now and mingling with the horde.
Add five (less one, a cow has balked and doubled on her track)
Hey! Slow 'em down! Somebody turn that bunch of 30 back!

Can I count up to 40? I bet the hauler thinks,
While they were sorting cattle, I was counting 40 winks.

Pairs: Cows with calves. Pot load: Semi, with trailer designed for hauling livestock. Alley: Narrow chute between corrals. Bred cows: Pregnant cows. Dry cows: Cows that are not pregnant. Yearling heifer: Young female, usually not bred. Brockle-face: Generally colored markings on a white face.

Living on the Land

Though you may not call the grasses by some scientific name,
But you know which kinds the cattle eat or pass by just the same,
And to see the pasture rich in good, sweet grass becomes your aim,
Then you're listening to and learning from the land.

If you know just where the great-horned owl is found as light grows dim,
And he doesn't fly because he knows you pose no threat to him,
When the deer don't flee and when the beaver stays and boldly swims,
You have forged some special friendships with the land.

When you search to find the blue-eyed grass that blossoms every year,
In a corner of the pasture that increasingly is dear
For the cycle of the seasons and the birdsong sweet and clear,
Then your roots are growing deeper in the land.

When you rise on winter mornings and go out before it's light,
And load up to feed the cattle, cold and hungry from the night,
And you pause to watch the sunrise, never tiring of the sight,
Then your soul expands and joins you to the land.

And the land returns your care through summer heat and winter snow,
From the shelter of the willows where the precious waters flow
To the dry and rocky hills where hardy native grasses grow.
You are whole when you become one with the land.

ANECDOTES

NO EXCUSE REQUIRED

Oak trees surrounded the meadow near the Visitor Center in Yosemite Valley. I was leading a group of about thirty folks of all ages. I pointed to the mistletoe hanging heavily in the branches and said, "If any of you have needed an excuse to kiss someone, there it is." A young girl—about seven years old—came out of the crowd and kissed me on the cheek. The entire group gave a simultaneous, warm-hearted, "Awwww…"

MOTHER NATURE SENDS SUBLIMINAL MESSAGES

I have no idea why I stopped along a trail one day and asked to group to step to the side, but they did, and a hawk suddenly soared down between us about two feet off the ground. This would have been a great moment to interpret birds of prey, but all I could think of was, "Wow!" It seemed to suffice.

THANK GOODNESS FOR GLITCHES

The Harpers Ferry Service Center in West Virginia sent a team out to Yosemite to videotape our programs. I was leading a group when a colorful but harmless mountain king snake slithered towards me. I kept backing up, but it was a speedy little rascal and I started bumping into rocks and shrubs as I tried to get out of its path. This undignified retreat had been filmed and I anticipated generations of NPS trainees laughing at the snake chasing the ranger. Then I heard a mild curse. The camera had malfunctioned, reducing the tape to a mangled, irretrievable mass. Phew!

BETTER THAN A DOG EATING HIS HOMEWORK

Somewhere out there is a man who, as a child, had the best possible excuse for losing his jacket in Yosemite. He was one of a group of children taking the Junior Ranger program and he had left his jacket on the outside of a row of seats. "A bear took it!" he told his parents. What parent would believe that? But I was able to corroborate his story. We had watched a young black bear grab it and disappear into the woods. It probably became a family tale to tell to skeptical listeners. He should have asked me for a sworn statement.

A FOUR-FOOTED FAREWELL

My last month in Yosemite was spent in the High Country after it was virtually shut down for the season. I strolled out onto a broad expanse of granite and leaned up against a boulder. My flat hat was on, and suddenly it wiggled. Then a Belding ground squirrel appeared over the brim, peeking upside-down at the startled and delighted ranger. It was a going-away gift.

THE RIGHT BAIT

Ranger hats make great frisbees. At Fort Point in Golden Gate National Recreation Area, the wind is nearly constant and I was standing on a pier, watching some fishermen when a sudden gust sent the hat sailing far out onto the choppy water of San Francisco Bay. The fishermen started casting for it and one "caught" it and reeled it in. Just for the record, his hook was baited with a tiny shrimp.

AN ARCH COMMENT

After 9/11, a homeland security detail at Bighorn Canyon National Recreation Area within the Crow Reservation introduced me to the hospitality and humor of the Crow Tribe. My duty was to guard the Yellowtail Dam, thirteen hours a night. That left me the daytime to get to know the country. One extremely hot day, the NPS maintenance crew offered me a boat ride out to a buoy where, I hoped, I'd get a brief respite from the 110-degree heat. Once underway, they added that they were going several miles upriver. I didn't object. They pointed out various features along the steep canyon walls. One was a stone arch. "The early settlers called that the Jug Handle," Marty informed me. "The Park Service renamed it 'Eye of the Eagle.'" I asked what the Crow People had called it. He shrugged. "Hole in a rock…"

MOOSE CALL FOR BACKUP

While doing a road patrol detail in Yellowstone to help with the influx of bikers during the famed Sturgis, South Dakota motorcycle rally, I pulled over a couple of bikers who had passed on a curve over a double yellow line. As I checked their licenses, four more bikers stopped and dismounted at one end of the pullout. Then another five stopped at the other end. I now had eleven bikers, all in their black leather gear, watching silently. Before I succumbed to the temptation to run for cover, a cow moose with a young calf came out of the woods and crossed the road into dense growth. The nine "extra" bikers rushed after the pair, I wrote the tickets in record time and the original two went on their way, probably laughing at my discomfiture. Good timing, moose.

A HOWLING SUCCESS

On a three-month detail to the Federal Law Enforcement Training Center in Georgia, I routinely jogged a six-mile loop with the full-time instructors and their wives. The conversation turned to wildlife and I mentioned how easy it was to call coyotes with a high-pitched howl. It was already after sunset and there didn't seem to be any lights around so I tipped back my head and howled. Oops. The very dense vegetation hid the fact that we weren't many yards from the town of Brunswick and my howl was picked up by dogs who passed it on until the entire town seemed to be in an uproar. It was reminiscent of Yosemite days when we would drive out to Mirror Lake and turn the patrol car siren on to "yelp" for a moment, and then listen to a coyote serenade.

HAROLD AND LUCY

Harold, a Hereford bull, was not happy. His harem of heifers had just been sold and his grumbles and anguished bellows could be heard all over the ranch. Meanwhile, Lucy, the goose, was loudly lamenting the loss of her mate, Christmas, who had been sent away for his habit of attacking ranch visitors. Gradually brought together by sorrow, they spent days commiserating, huge pink nose to orange bill. Alas, the romance was doomed. Harold found greater solace in hay and grain. Lucy, jilted, began attacking visitors and soon followed Christmas into exile. Shakespeare would have told the tale better.

CHILD'S PLAY

Sometimes I'd finish a demonstration for a school group (usually 4th graders) and need to stall for a few minutes before sending them on to the chuckwagon cook. I'd take them outside, have them make a circle and then say, "What if this was the only place you had to play and all you had to play with was your imagination. What would you do?" I loved the answers: Sing! Tell stories! Count all the rocks! Stand on our heads! Look for shapes in the clouds! There were so many answers, yet it almost never failed that one cheerful youngster would enthusiastically suggest, "Hit your brother!"

TAKING THE BULL BY THE HORNS

Well, to be honest, it was really a case of taking the cow by the tail. When a cow gets upset, as—for example—when it is being doctored in a chute, it tends to lash its tail around. A cow's tail is similar to a thick, hard rope, but your average rope isn't usually covered with fresh manure. Bill's dog, Spike, would wait by the chute, watching the tail like a snake charmer watches a cobra. When it writhed within range, he'd grab it, holding it firmly until the necessary branding and doctoring was done. Give that dog a big hand! And some mouthwash.

SPIKE AND BILL JOHNSON.

147

ABOUT THE AUTHOR

Lyndel was born in Helena, "in the first half of the last century," as she tells the high school students to whom she teaches basic blacksmithing at Powell County High School. She finds it's worth the price of divulging her antiquity as she watches the students work this out, their eyes widening in alarm. Charmingly, they then offer to lift heavy coal sacks for her until the horror fades and all returns to normal.

Her mother was born in California and her father in Ohio. After WWII they came to Montana to work a lead and silver mine six miles out of Elliston, west of Helena. It met with the usual success of such ventures and they moved to California, dragging their protesting youngster with them.

Over the years, Lyndel had three main ambitions: to be a cowgirl, a ranger, and go back to Montana. The circuitous path she followed to achieve all three of those goals is described in this book. But as she explains, it is really the story of "hands, hooves, claws, and paws." Her teachers ranged from several ants and caterpillar to university professors, curious children, and liars.

Her career took her from Vietnam in 1967 to Yosemite National Park, Fort Point, Alcatraz in San Francisco, and finally Grant-Kohrs Ranch National Historic Site in Deer Lodge, Montana. Along the way she discovered that a job could have many dimensions, and there was always something new to learn—and to challenge.

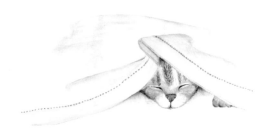